Effects of Early Nutrition on Premature Infants

Effects of Early Nutrition on Premature Infants

Guest Editor

Renato S. Procianoy

MDPI

Basel • Beijing • Wuhan • Barcelona • Belgrade • Novi Sad • Cluj • Manchester

Guest Editor
Renato S. Procianoy
Hospital de Clinicas de Porto
Alegre
Universidade Federal do Rio
Grande do Sul
Porto Alegre
Brazil

Editorial Office
MDPI AG
Grosspeteranlage 5
4052 Basel, Switzerland

This is a reprint of the Special Issue, published open access by the journal *Nutrients* (ISSN 2072-6643), freely accessible at: https://www.mdpi.com/journal/nutrients/special_issues/UMTAUQ731T.

For citation purposes, cite each article independently as indicated on the article page online and as indicated below:

Lastname, A.A.; Lastname, B.B. Article Title. *Journal Name* **Year**, *Volume Number*, Page Range.

ISBN 978-3-7258-4447-0 (Hbk)
ISBN 978-3-7258-4448-7 (PDF)
https://doi.org/10.3390/books978-3-7258-4448-7

Contents

About the Editor

Renato S. Procianoy

Renato S. Procianoy is a Full Professor of Pediatrics at Universidade Federal do Rio Grande do Sul in Porto Alegre, Brazil. He is a Full Member of the Brazilian Academy of Pediatrics and serves as the Chief Editor of *Jornal de Pediatria*. Additionally, he is a Member of the Neonatology Committee of the Brazilian Pediatric Society and a Researcher of the Brazilian Research Agency (CNPQ). His work focuses on pediatrics and neonatology, with a particular emphasis on nutrition and its impact on child development.

nutrients

MDPI

Editorial

Effects of Early Nutrition on Premature Infants

Rita C. Silveira and Renato S. Procianoy *

Hospital de Clínicas de Porto Alegre, Universidade Federal do Rio Grande do Sul, Porto Alegre 9045007, Brazil; drarita.c.s@gmail.com
* Correspondence: rprocianoy@gmail.com

Early nutrition plays a crucial role in both the short- and long-term health outcomes of premature infants, particularly those born with very low birth weight or extremely low gestational age. These infants are especially vulnerable due to their immature organs, limited nutrient reserves, and high metabolic demands [1,2].

A recent cohort study assessed the association between neonatal protein intake and brain structure at 7 years of age in children born very preterm, comparing two groups before and after a change in Neonatal Intensive Care Unit (NICU) nutritional protocol that increased protein intake. Although the group exposed to the higher-protein protocol showed reduced relative brain volume and cortical thinning in the occipital and parietal regions, absolute brain volumes were comparable between the groups. Higher neonatal intake of protein, fat, energy, and breast milk was associated with a more mature white matter microstructure, as indicated by higher fractional anisotropy and lower diffusivity on diffusion tensor imaging. These findings suggest that increased early protein intake may promote white matter maturation, with effects persisting into childhood [2].

In terms of nutritional recommendations for preterm neonates, human milk is the preferred form of enteral nutrition. The hierarchy of options includes the following: (1) raw mother's milk, (2) pasteurized mother's milk, and (3) pasteurized donor milk. Breastfeeding enhances mother–infant bonding and positively impacts neurodevelopment, highlighting the importance of family-centered care and breastfeeding support in NICUs [3].

Despite digestive immaturity, the early initiation of enteral feeding with human milk is strongly recommended—even for extremely preterm infants (<28 weeks' gestation or <1000 g birth weight). Delayed feeding, slow advancement, and routine gastric residual volume checks are outdated practices that prolong the need for parenteral nutrition and increase the risk of complications [4].

The fortification of human milk is essential for very low birth weight infants, as standard volumes of human milk alone may not meet their high nutritional requirements. Early fortification has been linked to improved growth trajectories without increasing the risk of metabolic disorders later in life. Studies show that higher protein fortification (≥1.4 g/100 mL) enhances growth without documented adverse effects. Individualized fortification—guided by an analysis of the mother's own milk (MOM) or pasteurized donor human milk (PDHM)—is a promising strategy to optimize nutrient intake in preterm infants. Although still emerging, evidence suggests benefits for clinical outcomes, particularly in cases of growth faltering.

Before conducting milk analysis, it is essential to rule out inadvertent fat losses during handling (e.g., milk transfers or prolonged pump feeding) and to ensure standard fortification has been fully optimized per local protocols. Milk analysis may then support targeted adjustments or prompt investigation of other causes of inadequate growth, such as increased metabolic demands, malabsorption, or genetic disorders. Clinicians should exercise caution when discussing nutrient analysis results with families to avoid undermining confidence in the adequacy of MOM. Accurate interpretation requires the use of true protein values (excluding non-protein nitrogen), metabolizable energy estimates, and representative samples. Currently, the lack of standardized methods for milk nutrient measurement and labeling limits comparability across settings, highlighting the need for further rigorous research [5,6].

Feeding strategies should include early colostrum expression (ideally within the first 6 h) and small, frequent doses administered via oral syringe. Standardized enteral feeding protocols help improve nutritional outcomes. Feed advancement should be based on birth weight and increased as tolerated. For extremely preterm infants, early fortification (starting on day 2) can improve linear growth and head circumference without increasing fat mass. Early feeding with human milk supports gastrointestinal development, helps establish a beneficial gut microbiome, and reduces the risk of necrotizing enterocolitis (NEC), a severe and potentially fatal intestinal condition [1,3]. Colostrum, rich in immunological components, plays a crucial role in early gut protection and immune system priming. Delayed or overly cautious advancement of feeds may result in nutritional deficits, poor growth, and suboptimal developmental outcomes.

Bolus feeding is generally preferred over continuous infusion, except in cases of intestinal dysmotility. Routine gastric residual checks should be avoided in favor of clinical assessment. High-volume feeding (>180 mL/kg/day) supports better growth and neurodevelopment, though special consideration is needed in conditions such as bronchopulmonary dysplasia or patent ductus arteriosus. Nutritional decisions should be guided by anthropometric monitoring based on gestational age and clinical comorbidities [1,3,4].

Early enteral nutrition—preferably with human milk—plays a pivotal role in reducing the need for parenteral nutrition (PN) in preterm infants, thereby minimizing associated risks. Early enteral feeding promotes gastrointestinal maturation, hormonal and metabolic adaptation, and supports neurodevelopment. The prompt initiation of human milk-based enteral nutrition can decrease the duration and volume of PN required, facilitating earlier removal of central lines and reducing the risk of central line-associated bloodstream infections, liver dysfunction, and sepsis [1,3,5].

Additionally, the gradual advancement of enteral feeds helps prevent refeeding syndrome—a metabolic disorder marked by electrolyte imbalances, particularly hypophosphatemia, that can occur when nutrition is reintroduced after undernutrition. Preterm infants, especially those $\leq$32 weeks' gestation, <1500 g, or with severe intrauterine growth restriction, are at elevated risk of refeeding syndrome during early PN due to limited nutrient stores and high metabolic demands. Thus, a balanced approach combining early enteral and PN—including the timely provision of amino acids and lipids and adequate electrolyte supplementation—is essential to ensure safe nutritional progression, metabolic stability, and optimal neurodevelopmental outcomes [1,2,4].

In conclusion, early nutrition in premature infants is more than a feeding strategy—it is a critical form of early intervention. Timely, appropriate, and individualized nutritional care can profoundly improve survival, growth, and developmental trajectories. The focus should be on early human milk-based enteral nutrition, with fortification when needed and minimal dependence on PN. Practices such as early colostrum expression and skin-to-skin care may improve breastfeeding rates and contribute to sustained developmental benefits and stronger mother–infant bonding.

Funding: This research received no external funding.

Institutional Review Board Statement: Not applicable.

Informed Consent Statement: Not applicable.

Data Availability Statement: Not applicable.

Conflicts of Interest: The authors declare no conflicts of interest.

References

1. Su, B.-H. Optimizing nutrition in preterm infants. *Pediatr. Neonatol.* **2014**, *55*, 5–13. [CrossRef] [PubMed]
2. Poppe, T.; Tottman, A.C.; Gamble, G.D.; Jiang, Y.; Silva, A.E.; Nguyen, L.; Harding, J.E.; Alsweiler, J.M.; Thompson, B.; PIANO study group. Neonatal Nutrition and Brain Structure at 7 Years in Children Born Very Preterm. *JAMA Netw. Open* **2025**, *8*, e2456080. [CrossRef] [PubMed]
3. Embleton, N.D.; Moltu, S.J.; Lapillonne, A.; van den Akker, C.H.P.; Carnielli, V.; Fusch, C.; Gerasimidis, K.; van Goudoever, J.B.; Haiden, N.; Iacobelli, S.; et al. Enteral Nutrition in Preterm Infants (2022): A Position Paper From the ESPGHAN Committee on Nutrition and Invited Experts. *J. Pediatr. Gastroenterol. Nutr.* **2023**, *76*, 248–268. [CrossRef] [PubMed]
4. Abiramalatha, T.; Thanigainathan, S.; Ramaswamy, V.V.; Rajaiah, B.; Ramakrishnan, S. Re-feeding versus discarding gastric residuals to improve growth in preterm infants. *Cochrane Database Syst. Rev.* **2023**, *6*, CD012940. [CrossRef] [PubMed]
5. Belfort, M.B.; Stellwagen, L.; North, K.; Unger, S.; O'Connor, D.L.; Perrin, M.T. Deciphering macronutrient information about human milk. *J. Perinatol.* **2024**, *44*, 1377–1381. [CrossRef] [PubMed]
6. Salas, A.A.; Gunawan, E.; Nguyen, K.; Reeves, A.; Argent, V.; Finck, A.; Carlo, W.A. Early Human Milk Fortification in Infants Born Extremely Preterm: A Randomized Trial. *Pediatrics* **2023**, *152*, e2023061603. [CrossRef] [PubMed]

nutrients

MDPI

Article

Preeclampsia and Future Implications on Growth and Body Composition in Preterm Infants

Elisabeth Calek [1], Julia Binder [2], Pilar Palmrich [2], Felix Eibensteiner [3], Alexandra Thajer [1], Karin Harreiter [1], Angelika Berger [1] and Christoph Binder [1,*]

[1] Department of Paediatrics and Adolescent Medicine, Division of Neonatology, Paediatric Intensive Care Medicine and Neuropaediatrics, Medical University of Vienna, 1090 Vienna, Austria; elisabeth.calek@meduniwien.ac.at (E.C.); alexandra.thajer@meduniwien.ac.at (A.T.); karin.harreiter@meduniwien.ac.at (K.H.); angelika.berger@meduniwien.ac.at (A.B.)

[2] Department of Obstetrics and Gynaecology, Medical University of Vienna, 1090 Vienna, Austria; julia.binder@meduniwien.ac.at (J.B.); pilar.palmrich@meduniwien.ac.at (P.P.)

[3] Department of Emergency Medicine, Medical University of Vienna, 1090 Vienna, Austria; felix.eibensteiner@meduniwien.ac.at

* Correspondence: christoph.a.binder@muv.ac.at; Tel.: +43-1-40400-67400

Abstract: Background: Preeclampsia is associated with intrauterine growth restriction (IUGR), which can lead to impaired postnatal growth and neurodevelopment in preterm infants. Preeclampsia can also occur without IUGR and its impact on postnatal nutrition, growth, and body composition remains not fully investigated to the best of our knowledge. Methods: This study included infants born before 37 weeks of gestation who underwent air displacement plethysmography to measure body composition (fat-free mass [FFM] and fat mass [FM]) at term-equivalent age. We compared infants born to mothers with preeclampsia and IUGR (PE-IUGR group) and preeclampsia without IUGR (PE-non-IUGR group) to those born to mothers without preeclampsia (control group). Results: In total, 291 infants were enrolled (control: $n = 227$; PE-non-IUGR: $n = 43$; PE-IUGR: $n = 21$). FFM was significantly lower in the PE-IUGR (mean differences −231 g (IQR: (−373, −88); $p < 0.001$)) and PE-non-IUGR groups (mean differences −260 g (IQR: (−372, −149); $p < 0.001$)) in comparison to the control group. FM was not significantly different between the three groups. Conclusions: This study indicates that infants of preeclamptic mothers, even without IUGR, had significantly lower FFM at term-equivalent age compared to the control group. Further research is necessary to determine if these variations can be modified.

Keywords: preeclampsia; intrauterine growth restriction; air displacement plethysmography; body composition; fat-free mass; nutrition; preterm infant

1. Introduction

Hypertensive disorders of pregnancy (HDP) occur in 5–10% of all pregnant women worldwide and are leading causes of maternal and fetal morbidity and mortality [1–3]. HDP is allocated in the following categories: gestational hypertension, chronic hypertension, preeclampsia, and preeclampsia superimposed upon chronic hypertension [4,5] Preeclampsia is a hypertensive disorder that typically involves the de novo onset of hypertension, including proteinuria, maternal organ failure, and/or uteroplacental dysfunction beyond 20 weeks of pregnancy [4]. A crucial differentiation in the diagnosis of preeclampsia is between early-onset preeclampsia (before 34 weeks of gestation) and late-onset preeclampsia (at or after 34 weeks of gestation) [6]. While there is some overlap in clinical features, early-onset preeclampsia typically presents with severe placental, maternal, and fetal clinical manifestations, resulting in adverse outcomes [7]. Despite ongoing research, there is still no consensus on the origin and pathogenesis of preeclampsia, and its pathophysiology remains not fully understood [8–10].

Preeclampsia is the most common serious form of hypertensive pregnancy complications, is linked to increased morbidity and mortality, and leads to a significant number of preterm birth and neonatal morbidity [7]. The long-term impact of preeclampsia impact both women and their newborns [6,7]. Preeclampsia frequently leads to preterm delivery, uteroplacental dysfunction, and prenatal growth restriction with increased rates of complications, like pulmonary hypertension, necrotizing enterocolitis, intraventricular hemorrhages, and, often, mortality, due to the combination of prematurity and extremely low birth weight [11,12]. However, preeclampsia can also occur without fetal growth restriction and with normal fetal doppler ultrasound parameters typically at later gestational ages after 34 weeks of gestation [13]. The etiology and pathomechanism in these cases are not fully investigated [14,15]. Furthermore, the long-term effect on the infant's health, especially concerning extrauterine growth and body composition, are unclear. Intrauterine growth restriction (IUGR) caused by preeclampsia is correlated with increased fetal and neonatal morbidity and mortality [16]. Ideal postnatal nutrition management and growth assessments for these infants have not been established. Postnatal growth faltering may be best defined by changes in z-scores [17]. However, enhanced neonatal growth, particularly gains in fat-free mass (FFM) measured by body composition, has been associated with brain size and improved neurodevelopment [18,19]. Furthermore, FFM is a good parameter to assess the nutritional status, whereas increased fat mass (FM) is connected with obesity and cardiovascular disorders [11,12].

As far as we know, the effects of preterm infants born to mothers with preeclampsia, particularly those without IUGR, on body composition remain unexplored. Consequently, this study aimed to assess how preeclampsia, both with and without IUGR, influences growth and body composition, particularly focusing on FFM at term-equivalent age.

2. Materials and Methods

2.1. Study Design and Setting

We conducted a retrospective cohort study that took place at the Department of Pediatrics and Adolescent Medicine, Division for Neonatology and Department of Obstetrics and Gynecology at the Medical University Vienna, Austria. Ethical approval was granted by the local Ethics Committee (Approval Number: 1602/2019).

The main objective of the study was to assess the impact of preeclampsia on growth and body composition in preterm infants. Various growth parameters, such as weight, length, and head circumference at both birth and term-equivalent age, were evaluated. Infants born to mothers with preeclampsia (PE group) were compared with those born to mothers without preeclampsia (control group). Additionally, the preeclampsia group was further subdivided into infants without intrauterine growth restriction (PE-non-IUGR group) and those with intrauterine growth restriction (PE-IUGR group). The definitions of these study groups are described below.

2.2. Population

This study comprised all preterm infants born before the 37th week of gestation who were admitted to the hospital between 2017 and 2023 and who had their body composition assessed. According to the local standard protocol, body composition measurements are routinely conducted for infants born preterm at term-equivalent age. Exclusion criteria included chromosomal abnormalities, as well as genetic and metabolic disorders.

Group assignment (control and PE groups) was based on the presence or absence of preeclampsia in the child's mother. Preeclampsia is determined as de novo onset of hypertension >20th week of gestation and at least one of the following conditions: proteinuria (≥300 mg/day), organ failure (renal dysfunction, hematological complications, such as thrombocytopenia, liver involvement, and neurological complications) or uteroplacental dysfunction, including growth restriction [6,20]. The PE group was further categorized into two subgroups: the PE-non-IUGR group and the PE-IUGR group. Allocation was based on the following criteria: fetal weight below the 10th percentile and evidence of placental insufficiency [21,22]. The diagnosis of IUGR was based on Gordjin et al. [21], which includes an estimated fetal weight < 10th percentile, a uterine artery and/or umbilical artery pulsatility index > 95th percentile, and/or a middle cerebral artery pulsatility index < 5th percentile. The three studies groups were defined in detail as follows: (1) The PE-non-IUGR group was defined as including infants from mothers with preeclampsia with normal Doppler ultrasound measurement and a fetal weight ≥ 10th percentile. (2) The PE-IUGR group was defined as including infants from mothers with preeclampsia with abnormal Doppler ultrasound measurement and a fetal weight < 10th percentile. (3) The control group was defined as infants born to mothers without preeclampsia and a birthweight appropriate for gestational age (between the ≥10th percentile and ≤90th percentile) [21].

Neonatal morbidity was defined as follows: Retinopathy of prematurity (ROP) [23], and intraventricular hemorrhage (IVH) characterized according to the criteria established by Papile et al. [24]. Bronchopulmonary dysplasia (BPD) has been identified as an oxygen demand > 21% at 36 plus 0 weeks of gestation [25]. Necrotizing enterocolitis (NEC) was defined according to the guidelines of Bell et al. [26]. Culture-proven sepsis was identified as a positive bacterial or fungal blood infection, accompanied by symptoms of infection, or antibiotic therapy for >5 days.

Population-based data comprised infants' age at delivery, sex, antenatal steroid drugs, preterm premature rupture of membranes (PPROM), pathological CTG (cardiotocography), prenatal infection-related preterm delivery and preeclampsia-related preterm delivery, mode of childbirth, APGAR score (5 as well as 10 min), umbilical artery pH, and birth weight, length, and head circumference measurements. Antihypertensive drugs including alpha methyldopa, urapidil, calcium channel blocker, and beta receptor blocker were analyzed. Alpha methyldopa is the local first-line therapy, and the following drugs were also used according to clinical condition and international guidelines: urapidil, calcium channel blocker, and beta receptor blocker [27].

2.3. *Measurements and Nutrition*

Weight, length, and head circumference measurements were conducted at birth, at discharge, and during the clinic follow-up visit when body composition was assessed. Daily weight was assessed and recorded every 48 h once the infant reached 1000 g. Body length was evaluated by a length panel, and head circumference by a flexible ruler.

Body composition was evaluated using air-displacement plethysmography (PEA POD® device; COSMED, Concord, CA, USA). The examination takes 5–7 min and measures fat-free mass (FFM) and fat mass (FM) [5]. It is a simple, reliable and, as it does not require anesthesia, is a safe method of measuring body composition [28,29].

FFM and FM variables were transformed into sex- and age-adjusted Z-Scores based on published reference data [29]. Fenton [30] and WHO growth charts were used for anthropometric data [31]. Anthropometric data were evaluated from childbirth to term age. Weight growth velocity (gram per kg per day) was calculated between day 7 and day 28 (average 2-point method).

Information on the type of feeding was collected during the inpatient stay. Early neonatal feeding began with breastmilk or, if unavailable, pasteurized preterm single donor milk (holder pasteurization) or nutrient-enriched formula for infants born preterm older than 32 weeks of gestation. This feeding regimen was implemented directly after birth and gradually increased between 20–30 milliliter per kg and day. Fortification was started at 100 milliliter per kg and day with Aptamil FMS, Nutricia, Frankfurt, Germany (infants with a gestational age > 26 weeks) or Humavant plus 6, Prolacta Bioscience, California, United States of America (infants <26 weeks). In addition, parenteral nutrition was imitated at birth, following the ESPGHAN guidelines for carbohydrate, protein, and fat intake [32]. Parenteral nutrition was stopped at 140–160 mL/kg and the day of enteral intake. At discharge, the enteral diet (including breastmilk, fortification, formula, and mixed feeds) was documented.

2.4. Statistics

Data analysis was conducted with SPSS version 28 (IBM, New York, NY, USA). A significance level of $p < 0.05$ was applied. Differences in baseline characteristics, growth velocity (measured in grams per kilogram per day from day 7 to day 28), nutrition at discharge (exclusively mother's own milk), and ROP, BPD, IVH, NEC, and sepsis were compared using Mann–Whitney U or Pearson's chi-square tests. The Mann–Whitney U and Pearson's chi-square tests were used accordantly, comparing the expected independent study groups and outcome parameters with the assumption that the data were not normal distributed.

Demographic details were displayed by frequency distribution, median, and interquartile range (IQR). To standardize measurements, anthropometric data and FFM/FM Z-scores were determined using growth charts adjusted for age and sex [33]. Multivariable regression analysis was applied to examine the relationship between FFM and FM. The FFM index (FFMI) and FM index (FMI) were calculated as follows: FM and FFM/lenght2 (kg/m^2), and weight at term age in the three study groups (control, PE-non-IUGR, PE-IUGR). The model was adjusted for confounding factors, namely sex [34], age at birth, and age at body composition measurement [35].

3. Results

In this investigation, 291 infants were analyzed (control: $n = 227$; PE: $n = 64$ [non-IUGR: $n = 43$, IUGR: $n = 21$]). Initially, 334 preterm infants were included, although 43 infants in the control group were rejected due to the following factors: large for gestational age: $n = 6$; SGA: $n = 34$; no body composition measurement due to continuous oxygen requirements: $n = 1$; and incomplete follow-up: $n = 2$ (Figure 1).

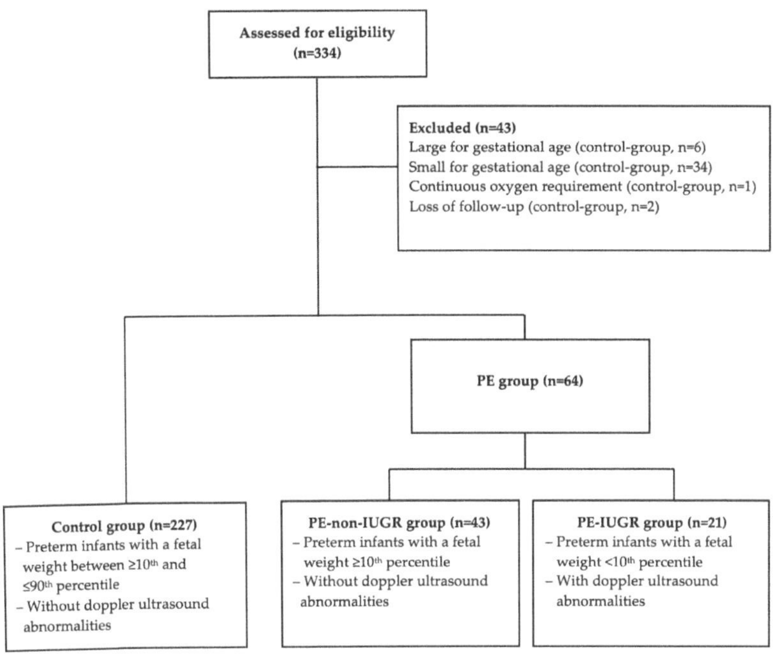

Figure 1. Overview of the study groups.

Baseline characteristics are presented in Table 1. At birth, median gestational age was not significantly different between the study groups ($p = 0.36$). Study groups did not differ significantly in proportion of females or males (control group: 59% male, PE-non-UGR group: 51% male, PE-IUGR group: 48% male) ($p = 0.86$). Anthropometric data were significantly lower at birth in infants in the PE groups compared to the control group ($p < 0.001$), (Table 1).

Table 1. Baseline characteristics.

Variables	Control Group (AGA-Group) (*n* = 227)	PE-Non-IUGR Group (*n* = 43)	PE IUGR Group (*n* = 21)	*p*-Values
Gestational age, weeks *	26.4 (25.0; 28.0)	27.6 (23.6; 32.1)	26.6 (25.1; 28.2)	0.36
Male, % (n)	59 (133/227)	51 (22/43)	48 (10/21)	0.86
Antenatal steroids, % (n)	84 (190/227)	88 (38/43)	86 (18/21)	0.73
PPROM, % (n)	46 (104/227)	7 (3/43)	5 (1/21)	<0.001
Pathological CTG, % (n)	17 (39/227)	28 (12/43)	38 (8/21)	0.03
Infection-related delivery, % (n)	73 (165/227)	7 (3/43)	5 (1/21)	<0.001
PE-related delivery, % (n)	0 (0/0)	93 (40/43)	95 (20/21)	<0.001
Antihypertensive drugs, % (n)				
- Alpha methyldopa	0 (0/0)	100 (43/43)	100 (21/21)	<0.001
- Drug combination **	0 (0/0)	23 (10/43)	28 (6/21)	<0.001
Caesarean delivery, % (n)	68 (155/227)	70 (30/43)	71 (15/21)	0.005
APGAR Score, 5 min *	9 (8, 9)	9 (8, 9)	9 (8/9)	0.40
APGAR Score, 10 min *	9 (9, 9)	9 (9, 9)	9 (9/9)	0.019
Umbilical artery, pH *	7.34 (7.28, 7.38)	7.31 (7.29, 7.33)	7.30 (7.27, 7.32)	0.024
Birth weight, gram *	910 (740, 1185)	820 (630, 1522)	600 (505, 850)	<0.001
Birth weight, Z-Score *	0.1 (−0.4, 0.5)	−0.7 (−1.1, −0.8)	−1.5 (−1.6, −1.3)	<0.001
Birth length, cm *	35 (32, 38)	35 (31, 41)	30 (29, 35)	<0.001
Birth length, Z-Score *	0.1 (−0.5, 0.8)	−0.5 (−0.9, −0.2)	−1.5 (−1.7, −1.4)	<0.001
Birth HC, cm *	25.0 (23.0, 26.5)	25.0 (21.0, 28.0)	22.0 (21.2, 25.2)	<0.001
Birth HC, Z-Score *	0.3 (−0.3, 1.0)	−0.5 (−0.8, −0.3)	−1.5 (−1.7, −1.4)	<0.001

* Data presented in median (interquartile range); preeclampsia (PE), preterm premature rupture of the membrane. (PPROM), head circumference (HC), cardiotocography (CTG). ** Additional antihypertensive drug therapy: urapidil, calcium channel blocker, or beta receptor blocker.

Causes of preterm delivery were significantly different between the groups (Table 1). Infection-related preterm delivery was the main reason in the control group and preeclampsia-related preterm delivery was the main reason in the PE groups. Antihypertensive drug therapies in the PE groups are displayed in Table 1. Preeclampsia occurred at a median gestational age of 22.4 weeks (IQR: 20.4; 24.6) in the PE-non-IUGR group and 22.3 weeks (IQR: 20.3; 24.1) in the PE-IUGR group ($p = 0.68$). APGAR score at 10 min and umbilical artery pH were significantly lower in the PE groups compared to the control group (Table 1).

Short-term outcome parameters and nutrition at discharge are detailed in Table 2. Outcome parameters did not show significant differences among the study groups (Table 2). At discharge, infants in all three study groups received primarily their mother's own milk, and no statistically significant differences were found between the nutritional diet (control vs. PE-non-IUGR, $p = 0.10$, and control vs. PE-IUGR, $p = 0.27$).

Table 3 presents the non-adjusted anthropometric data and FFM as well as FM at term age.

Median gestational age at the time of body composition measurements showed no significant differences between the groups: control 42.1 weeks (IQR: 40.1; 46.3), PE-IUGR 41.0 weeks (IQR: 39.0; 44.6), and PE-non-IUGR 41.0 weeks (40.0; 44.6); control versus PE-IUGR ($p = 0.42$) and control versus PE-non-IUGR ($p = 0.21$). At term, weight was significantly lower in the PE groups compared to the control group. Both FFM and FM grams were also lower in the PE groups versus controls. The other growth parameters were very similar across groups.

Growth parameters are presented in Table 4. At discharge, median gestational age was not significantly different between the groups: control 38.1 weeks (IQR: 37.0; 40.0),

PE-IUGR 38.7 weeks (IQR: 37.4; 39.3) and PE-non-IUGR 38.0 weeks (IQR: 37.0; 39.4); control versus PE-IUGR (p = 0.75) and control versus PE-non-IUGR (p = 0.29).

Table 2. Outcome parameters and nutrition at discharge.

	Control Group (n = 227)	PE-Non-IUGR Group (n = 43)	PE-IUGR Group (n = 21)	Control vs. Non-IUGR p-Values	Control vs. IUGR p-Values
Neonatal morbidities					
IVH (stage ≥ 3), % (n)	8 (19/227)	5 (2/43)	5 (1/21)	0.60	0.19
ROP (stage ≥ 3), % (n)	14 (31/227)	9 (4/43)	14 (3/21)	0.90	0.30
BPD, % (n)	11 (25/227)	9 (4/43)	10 (2/21)	0.17	0.27
NEC (stage ≥ 2), % (n)	7 (16/227)	5 (2/43)	5 (1/21)	0.60	0.17
Culture proven sepsis, % (n)	22 (49/227)	14 (6/43)	19 (4/21)	0.28	0.09
Nutrition at discharge					
Exclusive mother's own milk at discharge, % (n)	54 (122/227)	67 (29/43)	67 (14/21)	0.10	0.27

Values are median (interquartile range); intraventricular hemorrhage (IVH), retinopathy of prematurity (ROP), bronchopulmonary dysplasia (BPD), and necrotizing enterocolitis (NEC).

Table 3. Non-adjusted anthropometric parameters and body composition measurements.

Variables	Control Group (n = 227)	PE-Non-IUGR Group (n = 43)	PE-IUGR Group (n = 21)
Age at measurement, week	42.1 (40.1, 46.3)	41.0 (40.0, 446)	41.0 (39.0, 44.6)
Anthropometric parameters at term-equivalent age *			
Weight, gram	3590 (2944, 4470)	3168 (2731, 4272)	3024 (2605, 3650)
Length, cm	51.0 (49.0, 55.0)	53.0 (51.0, 55.0)	51.0 (48.0, 54.0)
Head circumference, cm	35.0 (34.0, 37.5)	37.5 (35.0, 38.2)	36.0 (33.0, 38.0)
Body composition parameters at term-equivalent age *			
FFM, percentage	78.3 (73.9, 83.6)	79.0 (74.3, 81.0)	78.5 (73.3, 82.6)
FM, percentage	21.7 (16.5, 26.1)	21.0 (19.0, 25.7)	21.5 (17.4, 26.7)
FFM, gram	2821 (2440, 3323)	2527 (2195, 3173)	2425 (2144, 2631)
FM, gram	769 (504, 1147)	641 (536, 1099)	599 (461, 1019)

* Data displayed are median (interquartile range).

Table 4. Growth parameters.

Variables	Control Group (n = 227)	PE-Non-IUGR Group (n = 43)	PE-IUGR Group (n = 21)	Control vs. Non-IUGR p-Values	Control vs. IUGR p-Values
Growth velocity from day 7 to day 28 *					
Age at discharge, week	38.1 (37.0, 40.0)	38.0 (37.0, 39.4)	38.7 (37.4, 39.3)	0.29	0.75
Discharge weight, gram	2785 (2433, 3075)	2750 (2425, 2970)	2462 (2150, 2790)	0.21	0.002
Weight velocity, g/kg/d	15.2 (12.5, 17.1)	14.6 (11.6, 16.3)	14.4 (11.9, 16.5)	0.26	0.24
Discharge length, cm	46.0 (44.0, 48.0)	47.0 (46.0, 49.0)	46.0 (44.5, 46.0)	0.011	0.08
Discharge HC, cm	32.6 (31.5, 33.9)	33.0 (32.0, 33.5)	32.0 (31.5, 32.0)	0.80	0.004

* Data shown in median (interquartile range).

Upon discharge, the PE-IUGR group had a significantly lower weight in comparison to the control group (p = 0.002). No significant weight difference was observed in the PE- non-IUGR versus controls (p = 0.21). In addition, the PE-non-IUGR group had a significantly greater length at discharge than the control group (p = 0.011). Length in infants in the PE-IUGR was very similar in comparison to infants in the control groups (p = 0.08). Weight velocity did not differ between the PE groups and control groups.

Regression analysis revealed that the primary outcome parameters, FFM Z-score and FFM grams, were significantly lower in infants in the PE-IUGR group compared to the control group (p = 0.008 and $p < 0.01$, respectively), and in infants in the PE-non-IUGR group than in the control group ($p = 0.002$ and $p < 0.001$, respectively) (Table 5). The FM Z-score was not significantly different between all study groups. FFMI was significantly lower in the PE-IUGR- and PE-non-IUGR groups ($p = 0.002$, $p < 0.001$, respectively) in comparison to the control group. FMI was not significantly different between the PE groups and the control group (Table 5). Weight at scan was significantly lower in the PE groups in comparison to the control group: control versus PE-IUGR ($p = 0.003$) and control versus PE-non-IUGR ($p < 0.001$) (Table 5).

Table 5. Weight and fat mass and fat-free mass data for control, PE-non-IUGR and PE-IUGR groups.

	Adjusted Mean			Adjusted Mean Difference	
	Control Group	**PE-Non-IUGR Group**	**PE-IUGR Group**	**PE-Non-IUGR Group**	**PE-IUGR Group**
Total (n)	227	43	21		
Weight at scan, gram [1]	3836 (3775, 3897)	3483 (3337, 3630)	3519 (3336, 3702)	−353 (−512, −193) $p < 0.001$	−317 (−508, −126) $p = 0.003$
FFM, Z-score	−1.0 (−1.2, −0.9)	−1.6 (−1.9, −1.4)	−1.5 (−1.8, −1.2)	−0.6 (−0.8, −0.5) $p = 0.002$	−0.5 (−0.7, −0.3) $p = 0.008$
FM, Z-score	1.0 (0.9, 1.1)	0.7 (0.4, 1.0)	0.7 (0.3, 1.1)	0.2 (0.1, 0.3) $p = 0.24$	0.1 (−0.1, 0.2) $p = 0.41$
FFM, gram	2959 (2919, 3000)	2699 (2597, 2801)	2728 (2592, 2866)	−260 (−372, −149) $p < 0.001$	−231 (−373, −88) $p < 0.001$
FM, gram	864 (824, 904)	781 (686, 876)	789 (660, 917)	83 (−21, 187) $p = 0.117$	75 (−75, 207) $p = 0.334$
FFM, Index	10.8 (10.6, 10.9)	9.9 (9.5, 10.3)	10.0 (9.5, 10.5)	−0.9 (−1.3, −0.6) $p < 0.001$	−0.8 (−1.4, −0.3) $p = 0.002$
FM, Index	3.0 (2.8, 3.1)	2.9 (2.6, 3.3)	2.7 (2.2, 3.2)	−0.1 (−0.4, 0.3) $p = 0.80$	−0.3 (−0.8, 1.5) $p = 0.18$

[1] Mean (95% CI) adjusted for sex, postmenstrual age at measurement, and age at birth.

4. Discussion

The study demonstrated that preterm infants born to mothers with preeclampsia had significantly different body composition at term age, particularly a reduction in FFM compared to infants born to mothers without preeclampsia. Furthermore, in a subgroup analysis, we found that infants from preeclamptic mothers without growth restriction during pregnancy had significantly lower FFM at term age than infants in the control group. This study highlights that preeclampsia affects body composition independently of IUGR. Therefore, the presence of preeclampsia should be particularly considered in the postnatal nutritional management of these infants. Emphasis should be placed on individualized postnatal nutritional strategies to address these specific growth and developmental challenges to avoid growth and long-term neurological impairment.

Preeclampsia is a significant cause of preterm birth and is also linked to uteroplacental dysfunction, which is linked to IUGR [6]. Consequently, women with preeclampsia are at higher risk for delivering infants with a low birth weight [36]. Early diagnosis, consistent fetal monitoring, and optimal postnatal management are essential to reduce neonatal and perinatal mortality and morbidity [12]. However, adequate postnatal nutritional management is among the most important preventive interventions [37,38]. Studies have shown that improved nutrition in these infants is essential for optimizing brain size and neurodevelopment [18,39]. In a previous study, we found that standard nutritional management according to ESPGHAN recommendations was insufficient for infants with intrauterine growth retardation [40]. Therefore, individualized nutritional strategies should be considered in these infants.

Growth is generally evaluated including weight, length, and head circumference; however, several studies have demonstrated that assessing qualitative growth by measuring body composition is more accurate for evaluating nutritional management and long-term outcomes [18,41]. In particular, optimizing FFM gain is important for reducing long-term cardiovascular comorbidities [42]. However, preeclampsia can also occur without IUGR, and the possible negative effect on infants' health, particularly on growth and body composition, is not well understood [14,43]. These infants typically receive standard nutritional management, but preeclampsia without IUGR may still negatively impact uteroplacental nutrient metabolism and extrauterine growth [15,44]. Consequently, the objective of the research was to investigate whether preeclampsia, without IUGR, has an effect on postnatal growth and body composition.

Our investigation revealed that infants in the PE groups had significantly lower body weight at birth and at term compared to control infants. Furthermore, FFM at term-equivalent age was significantly lower in preterm infants in the PE groups, regardless of the presence of IUGR. It is widely recognized that IUGR is linked to compromised postnatal growth, and our data are consistent with previous studies demonstrating that preeclampsia with IUGR negatively affects postnatal growth [45,46]. We also found that preeclampsia without IUGR impacts infant growth and body composition. The data are new and underline the hypothesis that preeclampsia without IUGR has an effect on postnatal growth. We hypothesized that preeclampsia without IUGR and doppler abnormalities might affect placental function and growth factors. In general, preterm infants experience an early separation from the placenta, leading to a premature disruption of growth factors (GF) and hormones [47,48]. The absence of these factors, such as placental growth hormone, human placental lactogen, maternal insulin-like GF 1 and 2, corticotropin-releasing hormone, leptin, insulin and thyroid hormones, can result in growth faltering and suboptimal body composition [48]. The metabolic endocrine disorder associated with preeclampsia is often unrecognized. Supplementation of the growth-stimulating hormones, such as insulin-like GF 1 and thyroid hormones, in infants born preterm, may help to support reduced growth faltering [48]. However, preeclampsia is additionally linked to a reduction in growth factors and hypoalbuminemia, both of which are associated with poor growth and may negatively affect extrauterine growth in these infants [49–51]. Furthermore, Roberts et al. [52] have shown that vascular remodeling in the arteries of women with PE is substantially different from in those without PE. These factors can substantially influence postnatal growth and neurodevelopment. Additionally, infant nutrition, particularly the composition of breast milk in mothers with preeclampsia, may affect postnatal growth. Previous studies [53] have demonstrated that lipid metabolism and lactogenesis are impaired in women with preeclampsia, which could influence breast milk composition and infant growth. An exploratory study by Beser et al. [54] showed that macronutrients in colostrum were not affected by preeclampsia. However, the analysis of macronutrients at different stages of lactation has not yet been investigated, and research with a larger cohort is needed. Our investigation underlines the hypothesis that these infants are at increased risk for growth faltering, and an adequate nutritional management is essential for this specific patient group. IUGR is associated with poor neurodevelopment, cardiovascular disease, reduced lung function, renal impairment, increased insulin resistance, and metabolic syndrome [55–57]. Studies have demonstrated that high FM, as measured by body composition, is linked to being overweight and cardiovascular diseases [47,57]. We also assessed FM in the study groups but failed to show differences in FM Z-scores between the PE groups and controls. This indicates that the PE groups might not be at an increased risk of developing obesity later in life.

After birth, nutritional management is particularly challenging for growth-retarded children. Infants with IUGR frequently need extended periods of parenteral nutrition, as establishing enteral nutrition is slower because of restricted food tolerance and reduced nutrient reserves and risk of complications, such as necrotizing enterocolitis [58]. Early aggressive nutritional management is attempted to counteract appropriate growth [59]. Preterm infants should receive fortified breastmilk or formula at least until term-equivalent age to ensure optimal growth [32]. Based on recommendation by European Society for Pediatric Gastroenterology, Hepatology, and Nutrition (ESPGHAN), fortification is still suggested at discharge, if an infant does not exceed the 10th percentile [32]. In our study, nutrition was managed according to ESPGHAN guidelines but was not adequate to prevent FFM loss in preterm infants in the PE groups, regardless of whether or not the fetus as growth-restricted.

These data highlight the necessity of early individualized nutritional management in infants born from preeclamptic mothers with intrauterine growth restriction to prevent malnutrition and poor postnatal growth [17]. Overall, breast milk is the ideal feeding option for these infants because of its positive impact on cardiovascular, neurological, and growth outcomes. Pasteurized donor milk is the best alternative choice [42]. In recent years, attention has focused on optimizing postnatal growth through nutritional interventions [60]. A study by Perrin et al. [61] showed that premature infants benefit from an individually adapted diet. Targeted fortification to optimize protein and energy content would be particularly beneficial for growth-retarded infants [62,63]. However, analyzing breast milk or human milk is very time-consuming and resource-demanding [61]. Further studies and randomized controlled trials are necessary to assess individualized nutritional management for these infants.

A key strength of this research is the qualitative monitoring of growth by body composition measurement, utilizing not only traditional anthropometric parameters, such as body weight, but also others. While weight gain and length growth alone are not optimal indicators of nutritional status [47], body composition measurements provide both quantitative and qualitative information on growth, distinguishing between FM and FFM [64]. Monitoring body composition is, thus, a fundamental part of improving nutritional outcomes [18,41,65]. FFM, in particular, is associated with brain size and serves as a good marker of neurodevelopmental outcomes [18,41]. Consequently, routine measurements could be a good and easy method to assess the nutritional status as well as subsequent neurodevelopment. Furthermore, it is an easy and safe approach for investigating adequate growth in preterm infants. Further studies are required to investigate future clinical and research implications.

Two weaknesses of the current study are the comparatively small cohort and its retrospective character. However, the number of infants in the PE group ($n = 64$) is relatively large for this population, providing new insights into the effects of preeclampsia without IUGR on infants' growth and body composition. Further research is needed to improve insight in the pathomechanism and long-term effects of preeclampsia without IUGR on the health of these preterm infants.

Our study found that FFM is significantly lower in the PE group, even in infants born to mothers with preeclampsia without IUGR. In comparison to our previous study, the body composition measurements of the PE groups showed very similar FFM Z-scores (PE-non-IUGR FFM Z-score −1.6 and PE-IUGR FFM Z-score −1.5) [40]. The FFM Z-scores of the control group in our previous and the current studies were consistent (FFM Z-score −1.1) [40]. Additionally, FFMI was calculated and found to be significantly lower in the PE groups compared to the control group. These data emphasize the importance of individualized and enhanced nutrition in these infants to improve long-term health.

5. Conclusions

A significant decrease in FFM was observed in infants born to mothers with preeclampsia compared to those born to mothers without preeclampsia. Even infants born to preeclamptic mothers who did not experience growth restriction during pregnancy had significantly lower FFM and body weight at term-equivalent age. This finding suggests that infants born to mothers with preeclampsia have altered body composition. Therefore, research is needed to understand whether these differences are modifiable.

Author Contributions: Conceptualization, C.B.; methodology, C.B. and E.C.; validation, E.C., J.B., A.T., K.H. and C.B.; formal analysis E.C. and C.B.; investigation E.C., K.H. and C.B.; resources, E.C., J.B., A.T., K.H., A.B. and C.B.; data curation E.C., F.E. and C.B.; writing—original draft preparation, E.C.; writing—review and editing E.C., J.B., P.P., F.E., A.T., K.H., A.B. and C.B.; visualization, E.C. and C.B.; supervision, C.B. All authors have read and agreed to the published version of the manuscript.

Funding: This research received no external funding.

Institutional Review Board Statement: The research was carried out in compliance with the Declaration of Helsinki and obtained approval (Ethics Committee, Medical University Vienna (Number 1602/2019, 2 July 2019).

Informed Consent Statement: No applicable.

Data Availability Statement: Data used in the study are available on request (corresponding author). The data are not publicly available because of studies in progress.

Conflicts of Interest: A.B. reports research support (Abbvie, Chiesi, Pfizer, Danone, and Nestle). C.B. obtained funding from Nestle, Danone, and Prolacta. This funding was not related to the current study. All other authors state no conflicts of interest.

References

1. Rana, S.; Lemoine, E.; Granger, J.P.; Karumanchi, S.A. Preeclampsia: Pathophysiology, Challenges, and Perspectives. *Circ. Res.* **2019**, *124*, 1094–1112. [CrossRef] [PubMed]
2. Wilkerson, R.G.; Ogunbodede, A.C. Hypertensive Disorders of Pregnancy. *Emerg. Med. Clin. N. Am.* **2019**, *37*, 301–316. [CrossRef] [PubMed]
3. Wu, P.; Green, M.; Myers, J.E. Hypertensive Disorders of Pregnancy. *BMJ* **2023**, *381*, e071653. [CrossRef] [PubMed]
4. Khedagi, A.M.; Bello, N.A. Hypertensive Disorders of Pregnancy. *Cardiol. Clin.* **2021**, *39*, 77–90. [CrossRef] [PubMed]
5. Sutton, A.L.M.; Harper, L.M.; Tita, A.T.N. Hypertensive Disorders in Pregnancy. *Obstet. Gynecol. Clin. N. Am.* **2018**, *45*, 333–347. [CrossRef]
6. Turbeville, H.R.; Sasser, J.M. Preeclampsia beyond Pregnancy: Long-Term Consequences for Mother and Child. *Am. J. Physiol.-Ren. Physiol.* **2020**, *318*, F1315–F1326. [CrossRef]
7. Bokslag, A.; Weissenbruch, M.V.; Mol, B.W.; de Preeclampsia Groot, C.J.M. Short and Long-Term Consequences for Mother and Neonate. *Early Hum. Dev.* **2016**, *102*, 47–50. [CrossRef]
8. Kalafat, E.; Thilaganathan, B. Cardiovascular Origins of Preeclampsia. *Curr. Opin. Obstet. Gynecol.* **2017**, *29*, 383–389. [CrossRef]
9. Melchiorre, K.; Sutherland, G.R.; Baltabaeva, A.; Liberati, M.; Thilaganathan, B. Maternal Cardiac Dysfunction and Remodeling in Women with Preeclampsia at Term. *Hypertension* **2011**, *57*, 85–93. [CrossRef]
10. Melchiorre, K.; Sharma, R.; Thilaganathan, B. Cardiovascular Implications in Preeclampsia: An Overview. *Circulation* **2014**, *130*, 703–714. [CrossRef]
11. Crispi, F.; Bijnens, B.; Figueras, F.; Bartrons, J.; Eixarch, E.; Noble, F.L.; Ahmed, A.; Gratacós, E. Fetal Growth Restriction Results in Remodeled and Less Efficient Hearts in Children. *Circulation* **2010**, *121*, 2427–2436. [CrossRef] [PubMed]
12. Demicheva, E.; Crispi, F. Long-Term Follow-Up of Intrauterine Growth Restriction: Cardiovascular Disorders. *Fetal Diagn. Ther.* **2014**, *36*, 143–153. [CrossRef] [PubMed]
13. Yagel, S.; Verlohren, S. Role of Placenta in Development of Pre-eclampsia: Revisited. *Ultrasound Obstet. Gynecol.* **2020**, *56*, 803–808. [CrossRef] [PubMed]
14. Milosevic-Stevanovic, J.; Krstic, M.; Radovic-Janosevic, D.; Stefanovic, M.; Antic, V.; Djordjevic, I. Preeclampsia with and without Intrauterine Growth Restriction–Two Pathogenetically Different Entities? *Hypertens. Pregnancy* **2016**, *35*, 573–582. [CrossRef] [PubMed]
15. Roberts, D.J.; Post, M.D. The Placenta in Pre-Eclampsia and Intrauterine Growth Restriction. *J. Clin. Pathol.* **2008**, *61*, 1254. [CrossRef]
16. Mazarico, E.; Meler, E.; Mendoza, M.; Herraiz, I.; Llurba, E.; Diego, R.D.; Comas, M.; Boada, D.; González, A.; Bonacina, E.; et al. Mortality and Severe Neurological Morbidity in Extremely Preterm Growth-restricted Fetuses. *Ultrasound Obstet. Gynecol.* **2023**, *62*, 788–795. [CrossRef]
17. Shah, P.S.; Wong, K.Y.; Merko, S.; Bishara, R.; Dunn, M.; Asztalos, E.; Darling, P.B. Postnatal Growth Failure in Preterm Infants: Ascertainment and Relation to Long-Term Outcome. *J. Périnat. Med.* **2006**, *34*, 484–489. [CrossRef]
18. Binder, C.; Buchmayer, J.; Thajer, A.; Giordano, V.; Schmidbauer, V.; Harreiter, K.; Klebermass-Schrehof, K.; Berger, A.; Goeral, K. Association between Fat-Free Mass and Brain Size in Extremely Preterm Infants. *Nutrients* **2021**, *13*, 4205. [CrossRef]
19. Ramel, S.E.; Gray, H.L.; Christiansen, E.; Boys, C.; Georgieff, M.K.; Demerath, E.W. Greater Early Gains in Fat-Free Mass, but Not Fat Mass, Are Associated with Improved Neurodevelopment at 1 Year Corrected Age for Prematurity in Very Low Birth Weight Preterm Infants. *J. Pediatr.* **2016**, *173*, 108–115. [CrossRef]
20. Tranquilli, A.L.; Dekker, G.; Magee, L.; Roberts, J.; Sibai, B.M.; Steyn, W.; Zeeman, G.G.; Brown, M.A. The Classification, Diagnosis and Management of the Hypertensive Disorders of Pregnancy: A Revised Statement from the ISSHP. *Pregnancy Hypertens. Int. J. Women's Cardiovasc. Health* **2014**, *4*, 97–104. [CrossRef]
21. Gordijn, S.J.; Beune, I.M.; Thilaganathan, B.; Papageorghiou, A.; Baschat, A.A.; Baker, P.N.; Silver, R.M.; Wynia, K.; Ganzevoort, W. Consensus Definition of Fetal Growth Restriction: A Delphi Procedure. *Ultrasound Obstet. Gynecol.* **2016**, *48*, 333–339. [CrossRef] [PubMed]

22. Khalil, A.; Beune, I.; Hecher, K.; Wynia, K.; Ganzevoort, W.; Reed, K.; Lewi, L.; Oepkes, D.; Gratacos, E.; Thilaganathan, B.; et al. Consensus Definition and Essential Reporting Parameters of Selective Fetal Growth Restriction in Twin Pregnancy: A Delphi Procedure. *Ultrasound Obstet. Gynecol.* **2019**, *53*, 47–54. [CrossRef] [PubMed]
23. International Committee for the Classification of Retinopathy of Prematurity. The International Classification of Retinopathy of Prematurity Revisited. *Arch. Ophthalmol.* **2005**, *123*, 991–999. [CrossRef]
24. Papile, L.-A.; Burstein, J.; Burstein, R.; Koffler, H. Incidence and Evolution of Subependymal and Intraventricular Hemorrhage: A Study of Infants with Birth Weights Less than 1500 Gm. *J. Pediatr.* **1978**, *92*, 529–534. [CrossRef]
25. Jobe, A.H.; Ncalari, E. Bronchopulmonary Dysplasia. *Am. J. Respir. Crit. Care Med.* **2012**, *163*, 1723–1729. [CrossRef]
26. Walsh, M.C.; Kliegman, R.M. Necrotizing Enterocolitis: Treatment Based on Staging Criteria. *Pediatr. Clin. N. Am.* **1986**, *33*, 179–201. [CrossRef] [PubMed]
27. Stepan, H.; Kuse-Föhl, S.; Klockenbusch, W.; Rath, W.; Schauf, B.; Walther, T.; Schlembach, D. Diagnosis and Treatment of Hypertensive Pregnancy Disorders. Guideline of DGGG (S1-Level, AWMF Registry No. 015/018, December 2013). *Geburtshilfe Frauenheilkd.* **2015**, *75*, 900–914. [CrossRef]
28. Ma, G.; Yao, M.; Liu, Y.; Lin, A.; Zou, H.; Urlando, A.; Wong, W.W.; Nommsen-Rivers, L.; Dewey, K.G. Validation of a New Pediatric Air-Displacement Plethysmograph for Assessing Body Composition in Infants. *Am. J. Clin. Nutr.* **2004**, *79*, 653–660. [CrossRef]
29. Norris, T.; Ramel, S.E.; Catalano, P.; Caoimh, C.N.; Roggero, P.; Murray, D.; Fields, D.A.; Demerath, E.W.; Johnson, W. New Charts for the Assessment of Body Composition, According to Air-Displacement Plethysmography, at Birth and across the First 6 Mo of Life. *Am. J. Clin. Nutr.* **2019**, *109*, 1353–1360. [CrossRef]
30. Fenton, T.R. A New Growth Chart for Preterm Babies: Babson and Benda's Chart Updated with Recent Data and a New Format. *BMC Pediatr.* **2003**, *3*, 13. [CrossRef]
31. WHO Multicentre Growth Reference Study Group. WHO Child Growth Standards Based on Length/Height, Weight and Age. *Acta Pædiatrica* **2006**, *95*, 76–85. [CrossRef] [PubMed]
32. Embleton, N.D.; Moltu, S.J.; Lapillonne, A.; van den Akker, C.H.P.; Carnielli, V.; Fusch, C.; Gerasimidis, K.; van Goudoever, J.B.; Haiden, N.; Iacobelli, S.; et al. Enteral Nutrition in Preterm Infants (2022): A Position Paper From the ESPGHAN Committee on Nutrition and Invited Experts. *J. Pediatr. Gastroenterol. Nutr.* **2023**, *76*, 248–268. [CrossRef]
33. Bahlmann, F.; Reinhard, I.; Krummenauer, F.; Neubert, S.; Macchiella, D.; Wellek, S. Blood Flow Velocity Waveforms of the Fetal Middle Cerebral Artery in a Normal Population: Reference Values from 18 Weeks to 42 Weeks of Gestation. *J. Périnat. Med.* **2002**, *30*, 490–501. [CrossRef]
34. Gale, C.; Logan, K.M.; Jeffries, S.; Parkinson, J.R.C.; Santhakumaran, S.; Uthaya, S.; Durighel, G.; Alavi, A.; Thomas, E.L.; Bell, J.D.; et al. Sexual Dimorphism in Relation to Adipose Tissue and Intrahepatocellular Lipid Deposition in Early Infancy. *Int. J. Obes.* **2015**, *39*, 629–632. [CrossRef] [PubMed]
35. Uthaya, S.; Thomas, E.L.; Hamilton, G.; Doré, C.J.; Bell, J.; Modi, N. Altered Adiposity after Extremely Preterm Birth. *Pediatr. Res.* **2005**, *57*, 211–215. [CrossRef]
36. Ødegård, R.A.; Vatten, L.J.; Nilsen, S.T.; Salvesen, K.Å.; Austgulen, R. Preeclampsia and Fetal Growth. *Obstet. Gynecol.* **2000**, *96*, 950–955. [CrossRef] [PubMed]
37. Sharma, D.; Shastri, S.; Farahbakhsh, N.; Sharma, P. Intrauterine Growth Restriction—Part 1. *J. Matern.-Fetal Neonatal Med.* **2016**, *29*, 3977–3987. [CrossRef]
38. Villar, J.; Belizan, J.M. The Timing Factor in the Pathophysiology of the Intrauterine Growth Retardation Syndrome. *Obstet. Gynecol. Surv.* **1982**, *37*, 499–506. [CrossRef] [PubMed]
39. Kosmeri, C.; Giapros, V.; Rallis, D.; Balomenou, F.; Serbis, A.; Baltogianni, M. Classification and Special Nutritional Needs of SGA Infants and Neonates of Multiple Pregnancies. *Nutrients* **2023**, *15*, 2736. [CrossRef]
40. Calek, E.; Binder, J.; Palmrich, P.; Eibensteiner, F.; Thajer, A.; Kainz, T.; Harreiter, K.; Berger, A.; Binder, C. Effects of Intrauterine Growth Restriction (IUGR) on Growth and Body Composition Compared to Constitutionally Small Infants. *Nutrients* **2023**, *15*, 4158. [CrossRef]
41. Pfister, K.M.; Zhang, L.; Miller, N.C.; Ingolfsland, E.C.; Demerath, E.W.; Ramel, S.E. Early Body Composition Changes Are Associated with Neurodevelopmental and Metabolic Outcomes at 4 Years of Age in Very Preterm Infants. *Pediatr. Res.* **2018**, *84*, 713–718. [CrossRef] [PubMed]
42. Kumar, R.K.; Singhal, A.; Vaidya, U.; Banerjee, S.; Anwar, F.; Rao, S. Optimizing Nutrition in Preterm Low Birth Weight Infants—Consensus Summary. *Front. Nutr.* **2017**, *4*, 20. [CrossRef] [PubMed]
43. Srinivas, S.K.; Edlow, A.G.; Neff, P.M.; Sammel, M.D.; Andrela, C.M.; Elovitz, M.A. Rethinking IUGR in Preeclampsia: Dependent or Independent of Maternal Hypertension? *J. Perinatol.* **2009**, *29*, 680–684. [CrossRef] [PubMed]
44. Dimitriadis, E.; Rolnik, D.L.; Zhou, W.; Estrada-Gutierrez, G.; Koga, K.; Francisco, R.P.V.; Whitehead, C.; Hyett, J.; Costa, F.d.S.; Nicolaides, K.; et al. Pre-Eclampsia. *Nat. Rev. Dis. Primers* **2023**, *9*, 8. [CrossRef] [PubMed]
45. Gidi, N.W.; Goldenberg, R.L.; Nigussie, A.K.; McClure, E.; Mekasha, A.; Worku, B.; Siebeck, M.; Genzel-Boroviczeny, O.; Muhe, L.M. Incidence and Associated Factors of Extrauterine Growth Restriction (EUGR) in Preterm Infants, a Cross-Sectional Study in Selected NICUs in Ethiopia. *BMJ Paediatr. Open* **2020**, *4*, e000765. [CrossRef]
46. Clark, R.H.; Thomas, P.; Peabody, J. Extrauterine Growth Restriction Remains a Serious Problem in Prematurely Born Neonates. *Pediatrics* **2003**, *111*, 986–990. [CrossRef]

47. Casirati, A.; Somaschini, A.; Perrone, M.; Vandoni, G.; Sebastiani, F.; Montagna, E.; Somaschini, M.; Caccialanza, R. Preterm Birth and Metabolic Implications on Later Life: A Narrative Review Focused on Body Composition. *Front. Nutr.* **2022**, *9*, 978271. [CrossRef]
48. Möllers, L.S.; Yousuf, E.I.; Hamatschek, C.; Morrison, K.M.; Hermanussen, M.; Fusch, C.; Rochow, N. Metabolic-Endocrine Disruption Due to Preterm Birth Impacts Growth, Body Composition, and Neonatal Outcome. *Pediatr. Res.* **2022**, *91*, 1350–1360. [CrossRef]
49. Torry, D.S.; Wang, H.-S.; Wang, T.-H.; Caudle, M.R.; Torry, R.J. Preeclampsia Is Associated with Reduced Serum Levels of Placenta Growth Factor. *Am. J. Obstet. Gynecol.* **1998**, *179*, 1539–1544. [CrossRef]
50. Kimura, C.; Watanabe, K.; Iwasaki, A.; Mori, T.; Matsushita, H.; Shinohara, K.; Wakatsuki, A. The Severity of Hypoxic Changes and Oxidative DNA Damage in the Placenta of Early-Onset Preeclamptic Women and Fetal Growth Restriction. *J. Matern.-Fetal Neonatal Med.* **2013**, *26*, 491–496. [CrossRef]
51. Saitou, T.; Watanabe, K.; Kinoshita, H.; Iwasaki, A.; Owaki, Y.; Matsushita, H.; Wakatsuki, A. Hypoalbuminemia Is Related to Endothelial Dysfunction Resulting from Oxidative Stress in Parturients with Preeclampsia. *Nagoya J. Méd. Sci.* **2021**, *83*, 741–748. [CrossRef] [PubMed]
52. Roberts, J.M.; Escudero, C. The Placenta in Preeclampsia. *Pregnancy Hypertens. Int. J. Women's Cardiovasc. Health* **2012**, *2*, 72–83. [CrossRef] [PubMed]
53. Stadler, J.T.; Scharnagl, H.; Wadsack, C.; Marsche, G. Preeclampsia Affects Lipid Metabolism and HDL Function in Mothers and Their Offspring. *Antioxidants* **2023**, *12*, 795. [CrossRef] [PubMed]
54. Beser, E.; Cetinkaya, A.K.; Keser, M.K.; Okman, E.; Sari, F.N.; Dizdar, E.A. Evaluation of Breast Milk Macronutrient Content in Preeclamptic Mothers. *Breastfeed. Med.* **2022**, *17*, 318–321. [CrossRef] [PubMed]
55. Sacchi, C.; Marino, C.; Nosarti, C.; Vieno, A.; Visentin, S.; Simonelli, A. Association of Intrauterine Growth Restriction and Small for Gestational Age Status with Childhood Cognitive Outcomes. *JAMA Pediatr.* **2020**, *174*, 772–781. [CrossRef]
56. Gounaris, A.K.; Sokou, R.; Gounari, E.A.; Panagiotounakou, P.; Grivea, I.N. Extrauterine Growth Restriction and Optimal Growth of Very Preterm Neonates: State of the Art. *Nutrients* **2023**, *15*, 3231. [CrossRef]
57. Kesavan, K.; Devaskar, S.U. Intrauterine Growth Restriction Postnatal Monitoring and Outcomes. *Pediatr. Clin. N. Am.* **2019**, *66*, 403–423. [CrossRef]
58. Morgan, J.; Young, L.; McGuire, W. Delayed Introduction of Progressive Enteral Feeds to Prevent Necrotising Enterocolitis in Very Low Birth Weight Infants. *Cochrane Database Syst. Rev.* **2014**, CD001970. [CrossRef]
59. Ziegler, E.E.; Thureen, P.J.; Carlson, S.J. Aggressive Nutrition of the Very Low Birthweight Infant. *Clin. Perinatol.* **2002**, *29*, 225–244. [CrossRef]
60. Ramirez, C.B.; McCoy, K.L.; Jacob, R.; Lavender, E.; Bonagurio, K.; Guerra, D.A.; Karottakuttu, S.; Gelfond, J.; McCurnin, D.; Blanco, C.L.; et al. Effects of Human Milk on Body Composition and Growth in Very Low Birthweight Infants. *Pediatr. Res.* **2023**, *93*, 2028–2035. [CrossRef]
61. Perrin, T.; Pradat, P.; Larcade, J.; Masclef-Imbert, M.; Pastor-Diez, B.; Picaud, J.-C. Postnatal Growth and Body Composition in Extremely Low Birth Weight Infants Fed with Individually Adjusted Fortified Human Milk: A Cohort Study. *Eur. J. Pediatr.* **2023**, *182*, 1143–1154. [CrossRef] [PubMed]
62. Rochow, N.; Fusch, G.; Ali, A.; Bhatia, A.; So, H.Y.; Iskander, R.; Chessell, L.; Helou, S.E.; Fusch, C. Individualized Target Fortification of Breast Milk with Protein, Carbohydrates, and Fat for Preterm Infants: A Double-Blind Randomized Controlled Trial. *Clin. Nutr.* **2021**, *40*, 54–63. [CrossRef] [PubMed]
63. Cardoso, M.; Virella, D.; Papoila, A.L.; Alves, M.; Macedo, I.; Silva, D.E.; Pereira-da-Silva, L. Individualized Fortification Based on Measured Macronutrient Content of Human Milk Improves Growth and Body Composition in Infants Born Less than 33 Weeks: A Mixed-Cohort Study. *Nutrients* **2023**, *15*, 1533. [CrossRef] [PubMed]
64. Cerasani, J.; Ceroni, F.; Cosmi, V.D.; Mazzocchi, A.; Morniroli, D.; Roggero, P.; Mosca, F.; Agostoni, C.; Giannì, M.L. Human Milk Feeding and Preterm Infants' Growth and Body Composition: A Literature Review. *Nutrients* **2020**, *12*, 1155. [CrossRef]
65. Andrews, E.T.; Beattie, R.M.; Johnson, M.J. Measuring Body Composition in the Preterm Infant: Evidence Base and Practicalities. *Clin. Nutr.* **2019**, *38*, 2521–2530. [CrossRef]

nutrients

MDPI

Article

Impact of Refeeding Syndrome on the Short-Term Clinical Outcomes of Very-Premature Infants

Mountasser M. Al-Mouqdad [1,*], Belal Alshaikh [2], Haider H. Sumaily [1], Ameen A. Almotiri [1], Nabeel A. Alodhaidan [1], Latifah AlMahmoud [1], Adli Abdelrahim [1], Tamadur E. Yousif [1], Abdullah S. Alghamdi [1], Yasir A. Albarrak [3], Aljohara O. Alnafiey [4], Maha R. Al-Anazi [4], Thanaa M. Khalil [5], Raneem S. Asfour [6] and Suzan S. Asfour [7,*]

1 Neonatal Intensive Care, Hospital of Pediatrics, King Saud Medical City, Riyadh 12746, Saudi Arabia
2 Department of Pediatrics, Cumming School of Medicine, University of Calgary, Calgary, AB T2N 1N4, Canada
3 Pediatric Gastroenterology Department, King Saud Medical City, Riyadh 12746, Saudi Arabia
4 Pharmacy Department, Pharmaceutical Care Services, King Saud Medical City, Riyadh 12746, Saudi Arabia
5 Obstetric and Gynecology Department, Maternity Hospital, King Saud Medical City, Riyadh 12746, Saudi Arabia
6 Pharmacy College, Jordan University of Science and Technology, P.O. Box 3030, Irbid 22110, Jordan
7 Clinical Pharmacy Department, Pharmaceutical Care Services, King Saud Medical City, Riyadh 12746, Saudi Arabia
* Correspondence: m.almouqdad@ksmc.med.sa (M.M.A.-M.); asfsuzan@gmail.com (S.S.A.); Tel.: +96-6118371777 (M.M.A.-M. & S.S.A.); Fax: +96-6114354406 (M.M.A.-M. & S.S.A.)

Abstract: Background: Refeeding syndrome (RFS) is a potentially life-threatening condition that can occur in preterm infants if nutritional support is initiated or increased after a period of starvation or malnutrition. Objectives: The current study aimed to examine the short-term clinical outcomes of RFS in preterm infants born at ≤32 weeks of gestation. Methods: Infants with a gestational age of ≤32 weeks and a birth weight of <1500 g who were born and admitted to the level III neonatal intensive care unit and received parenteral nutrition upon admission were retrospectively evaluated. The modified log Poisson regression with generalized linear models and a robust variance estimator was applied to adjust the outcomes of infants. Results: In total, 760 infants met this study's inclusion criteria. Of them, 289 (38%) developed RFS. RFS was significantly associated with a composite outcome of mortality and intraventricular hemorrhage. Based on the multivariate Cox regression analysis adjusted for significant potential confounders, RFS was significantly associated with increased mortality risk, with a hazard ratio for death in infants with RFS being 1.74-fold higher compared to those without RFS. Conclusions: Preterm infants born at ≤32 weeks of gestation who develop RFS within the first week of life are at increased risk for both intraventricular hemorrhage and mortality. This study underscores the need for standardized clinical approaches for managing RFS in the neonatal intensive care unit to improve outcomes. Future research should establish a unified RFS definition and conduct clinical trials to optimize parenteral nutrition strategies for this vulnerable population.

Keywords: parenteral nutrition; refeeding syndrome; mortality; intraventricular hemorrhage; preterm infants

1. Introduction

Refeeding syndrome (RFS) occurs when nutrition is rapidly reintroduced after a period of prolonged starvation, leading to life-threatening shifts in fluids and electrolytes. This syndrome can cause serious metabolic disturbances, affecting the heart, lungs, blood, and nervous system [1]. The condition was first noted during World War II when individuals recovering from famine unexpectedly fell ill after receiving food. In 1951, Schnitker and colleagues reported that one-fifth of Japanese prisoners who had been starved in prison camps died suddenly after being re-fed and provided with vitamins [2].

During starvation, the body experiences depletion of key nutrients such as potassium and phosphorus [3,4]. When refeeding is started, increased insulin levels drive these electrolytes into the cells, leading to low blood phosphorus and potassium levels [5]. This can cause severe complications such as cardiac arrhythmias, muscle weakness, respiratory failure, convulsions, and encephalopathy. In particular, phosphorus depletion impairs energy production and oxygen delivery to tissues, which can be life-threatening [6,7].

Preterm infants who miss the crucial period of nutrient accumulation in the third trimester of pregnancy are born with insufficient nutrient reserves [8]. Infants who are small-for-gestational-age (SGA) or have intrauterine growth restriction (IUGR) are further at risk of nutrient deficiencies, which are often associated with placental insufficiency [9]. Adequate protein intake immediately after birth is considered essential for supporting growth and neurodevelopment by enhancing protein accretion and activating insulin-like growth factor-I pathways [10,11]. Health providers generally recommend that preterm infants should have a protein intake of 3.5–4.5 g/kg/day [12,13]. However, excessive protein intake can result in metabolic complications including acidosis, hyperammonemia, elevated blood urea nitrogen levels, and RFS [10,14,15]. Our study showed that the incidence rate of RFS in preterm infants born before 32 weeks of gestation is 38% [16]. The reported incidence of RFS in the literature varies widely. For example, in the ProVIDe trial, 20% of extremely low-birth-weight infants presented with RFS. Meanwhile, other studies have reported that the incidence rates of RFS were as high as 90%, particularly in SGA infants, those with IUGR, and those receiving aggressive parenteral nutrition—a high-energy nutrition plan from the first day of life [14,17–20]. This wide variability in the incidence rates may be attributed to differences in the definition of RFS, population characteristics, and sex. Despite this variability, RFS remains a significant and prevalent condition in preterm infants worldwide.

Several studies have explored the short-term neonatal outcomes associated with RFS, including intraventricular hemorrhage (IVH), bronchopulmonary dysplasia (BPD), and late-onset sepsis (LOS). However, the results are conflicting and inconclusive. Al-Wassia et al. and Cormack et al. found that premature infants who developed RFS had a higher incidence of IVH and severe IVH [14,17]. Conversely, other studies have found no significant association between RFS and IVH [9,19,21,22].

Ross et al. observed that RFS is associated with an increased risk of developing BPD in very-low-birth-weight (VLBW) infants. However, other studies reported no significant difference [9,14,19,21–23]. Similarly, Moltu et al. identified an association between severe hypophosphatemia and an increased risk of sepsis in VLBW infants. However, this finding was not supported by other studies, which did not find significant differences in the incidence of LOS [9,14,17,22–25].

Despite these findings, current studies are limited by several factors. Several studies have small sample sizes, and they included both preterm and term infants and used inconsistent definitions for RFS and hypophosphatemia [20,24,26,27]. In addition, these studies do not often report the concentrations of other parenteral nutrition (PN) components, such as dextrose and proteins [9,17,24]. Due to these limitations, it is challenging to accurately determine the actual clinical outcomes of RFS in preterm infants. Therefore, this study aimed to assess the short-term clinical outcomes of premature infants born at ≤32 weeks of gestation who developed RFS in their first week of life.

2. Materials and Methods

2.1. Study Design

This study retrospectively performed a review of medical documentation of preterm infants who were admitted to the neonatal intensive care unit (NICU) of King Saud Medical City (KSMC), a tertiary referral center, between January 2015 and June 2024. The average annual number of admissions in the level III NICU at KSMC is 1100.

This study was conducted in accordance with the Declaration of Helsinki and the Good Pharmacoepidemiology Practice Guidelines and was approved by the Medical Ethical Review Committee of KSMC (reference number: H1RI-12-May24-01). The need for informed consent was waived.

2.2. Inclusion and Exclusion Criteria

The inclusion criteria were as follows: very-preterm infants with a VLBW (<1500 g) who were born at KSMC, admitted to the level III NICU, and received PN plus lipid emulsion within the first 24 h of life.

The exclusion criteria were as follows: infants with known genetic or chromosomal abnormality, those with congenital infections or significant congenital defects, those who did not receive PN, those who were not born at KSMC or transferred to another hospital or died within the first 7 days after birth, and those who had nonretrievable data.

2.3. Data Collection and Follow-Up

The data of infants were collected from birth until death or discharge. The following data were obtained: demographic and clinical characteristics and outcomes, including major morbidities related to prematurity. Further, maternal data, including type of delivery, antenatal steroid treatment, and presence of gestational diabetes mellitus and maternal hypertension, were collected.

2.4. Study Outcome

The primary outcome of this study was the short-term clinical outcome of RFS in preterm infants born at $\leq$32 weeks of gestation.

2.5. Definitions

Nutrition Protocol

PN: Treatment with PN was started early after birth using starter PN. Individualized PN was prescribed daily. Starter PN contains 10% dextrose, 4% amino acids, and 0.01 mmol/mL of calcium gluconate [1]. Individualized PN solution containing amino acids (3.5–4 g/kg/day), dextrose (5–12 mg/kg/min), Lipid emulsion (1–3 g/kg/day), minerals, sodium chloride (1–3 mmol/kg/day), sodium acetate (1–2 mmol/kg/day), sodium phosphate (1–2 mmol/kg/day), potassium chloride (1–3 mmol/kg/day), potassium acetate (1–2 mmol/kg/day), potassium phosphate (1–2 mmol/kg/day), trace elements (Peditrace®), and water- and fat-soluble vitamins (Soluvit® N, and Vitalipid® N Infant; respectively) was started within the first 24 h of life and infused continuously for 24 h [1].

RFS: A clear definition of neonatal RFS has not yet been established. The following definitions were used: hypercalcemia, >2.8 mmol·L^{-1}; hypophosphatemia, >1.1 to <1.6 mmol·L^{-1}; and severe hypophosphatemia, <1.0 mmol·L^{-1} in the first week of life [14,17,27–29].

IVH: IVH was classified into grades I–IV according to the IVH classification of Papile et al. [30]. IVH was diagnosed based on the findings of head ultrasound performed between days 5 and 7 after birth [31,32]. All ultrasonography scans were performed by one expert radiologist and checked by another expert radiologist. 7.5- and 10-MHz transducers (LOGIQ e; GE Medical Systems Co., Ltd., Nanjing, China) were used to perform ultrasonography in sagittal and coronal planes.

2.6. Statistical Analysis

Before the analysis, the dataset was reviewed and checked for missing data. Data were analyzed using the Statistical Package for the Social Sciences software for Windows version 25.0 (IBM Corp., Armonk, NY, USA).

Infant and maternal variables were presented as descriptive statistics (median, interquartile range, frequency, and percentage). The Mann–Whitney U test was used for between-group comparisons of ordinal qualitative variables. The Fisher's exact test was utilized to determine the association between categorical variables. The unpaired Student's *t*-test was used for between-group comparisons of continuous variables with a normal distribution. The Mann–Whitney U test was used to assess variables with a non-normal distribution. The Kolmogorov–Smirnov test and a visual inspection of histograms were performed to evaluate the distribution of quantitative variables.

To determine the association between RFS in premature infants and neonatal outcomes, a univariate relative risk analysis of the recorded variables (gestational age, birth weight, SGA, delivery mode, sex, 1- and 5-min Apgar scores, maternal hypertension, antenatal steroid treatment, premature rupture of the membrane, gestational diabetes mellitus, necrotizing enterocolitis, surfactant use, LOS, dextrose intake, amino acids, and lipid emulsion, phosphate intake) was initially performed because the abovementioned factors were considered potential confounders. All factors with a p-value of <0.05 in the univariate analysis were included in the final multivariate regression analysis. The modified log Poisson regression with generalized linear models and a robust variance estimator (Huber–White) were applied in the univariate relative risk analysis and to the models to adjust the relative risk for the association between RFS and neonatal outcomes. All statistical tests were two-tailed, and p-values of <0.05 were considered significant.

3. Results

In total, 2465 preterm infants with a gestational age of ≤32 weeks and a birth weight of <1500 g were admitted to the level 3 NICU. Of them, 760 met the inclusion criteria and were eligible for the final analysis (Figure 1).

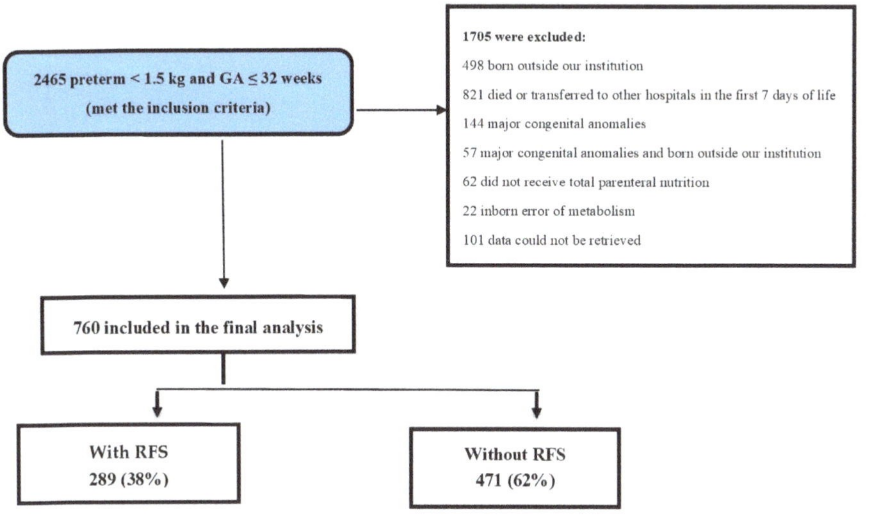

Figure 1. Flow chart of patient selection. *GA* gestational age.

Further, 289 (38%) of 760 infants developed RFS. Of 760 patients, 264 were aged <28 gestational weeks. Tables 1 and 2 show the demographic characteristics of the mothers and infants stratified according to RFS.

Table 1. Maternal characteristics of the participants stratified according to RFS.

Gestational Age	<32 Weeks				<28 Weeks			
Parameters	N	Infants without Refeeding Syndrome (n = 471)	Infants with Refeeding Syndrome (n = 289)	*p*-Value	N	Infants without Refeeding Syndrome (n = 128)	Infants with Refeeding Syndrome (n = 136)	*p*-Value
Antenatal steroid treatment, n (%)	760	253 (53.7)	147 (50.9)	0.45	264	70 (54.7)	71 (52.2)	0.71
Gestational diabetes mellitus, n (%)	760	27 (4.4)	14 (5.9)	0.37	264	7 (5.5)	4 (2.9)	0.36
Maternal hypertension, n (%)	760	119 (25.3)	67 (23.2)	0.54	264	26 (20.3)	27 (19.9)	1
Preterm rupture of membrane, n (%)	760	59 (13.1)	20 (7.2)	0.01 *	264	17 (13.3)	12 (8.8)	0.33
Cesarean section, n (%)	760	223 (47.3)	146 (50.5)	0.41	264	72 (56.3)	75 (55.1)	0.90

* *p*-values < 0.05.

Table 2. Neonatal characteristics of the participants stratified according to RFS.

Gestational Age	<32 Weeks				<28 Weeks			
Parameters	N	Infants without Refeeding Syndrome (n = 471)	Infants with Refeeding Syndrome (n = 289)	*p*-Value	N	Infants without Refeeding Syndrome (n = 128)	Infants with Refeeding Syndrome (n = 136)	*p*-Value
Gestational age (weeks), median (IQR)	760	29 (27.0–31.0)	28 (26–30.0)	<0.001 *	264	26 (25.0–27.0)	26 (25.0–27.0)	0.02 *
Birth weight (grams), (IQR)	760	1180 (950–1370)	950 (765–1200)	<0.001 *	264	845 (711.25–960)	775 (661.25–905)	0.008 *
Length (cm), median (IQR)	760	38 (35–40)	35 (32–38)	<0.001 *	264	33 (31–35)	33 (31–35)	0.10
Head circumference (cm), (IQR)	760	27 (25–28)	25 (23–27)	<0.001 *	264	24 (23–25)	23 (22–25)	0.03 *
1-min Apgar score, median (IQR)	760	6 (4–7)	5 (3–6)	<0.001 *	264	5 (3–6)	4 (2–6)	0.03 *
5-min Apgar score, median (IQR)	760	7 (6–8)	7 (7–8)	<0.001 *	264	5 (6–7)	4 (6–7)	0.17
Male sex, n (%)	760	221 (46.9)	174 (60.2)	<0.001 *	264	70 (54.7)	90 (66.2)	0.06
Expressed breast milk	760	247 (52.4)	145 (50.2)	0.55	264	75 (58.6)	64 (47.1)	0.06
Noninvasive respiratory support, n (%)	760	416 (88.3)	218 (75.4)	<0.001 *	264	96 (75)	79 (58.1)	0.004 *
Respiratory distress syndrome requiring surfactant, n (%)	760	281 (59.7)	224 (77.5)	<0.001 *	264	120 (93.8)	128 (94.1)	1
Patent ductus arteriosus requiring treatment, n (%)	760	35 (7.4)	34 (11.8)	0.05	264	24 (18.8)	28 (20.6)	0.76
Peripherally inserted central catheter (PICC), n (%)	760	154 (32.7)	121 (41.9)	0.01 *	264	60 (46.9)	75 (55.1)	0.22
Umbilical arterial catheter (UAC), n (%)	760	134 (28.5)	117 (40.5)	0.001 *	264	71 (55.5)	73 (53.7)	0.81
Umbilical venous catheter (UVC), n (%)	760	356 (75.6)	250 (86.5)	<0.001 *	264	121 (94.5)	125 (91.9)	0.47
Central venous catheter (CVC), n (%)	760	22 (4.7)	25 (8.7)	0.03 *	264	17 (13.3)	17 (12.5)	0.86
Average parenteral lipid intake within the first 7 days (g/kg/day), median (IQR)	760	2 (1.54–2.35)	2.1 (1.67–2.5)	0.02 *	264	1.8 (1.4–2.3)	1.8 (1.3–2.4)	0.95
Average parenteral protein intake within the first 7 days (g/kg/day), median (IQR)	760	3.90 (3.67–4.0)	3.97 (3.75–4.0)	0.01 *	264	4 (3.7–4.0)	4 (3.8–4.0)	0.13

Table 2. *Cont.*

Gestational Age	<32 Weeks				<28 Weeks			
Parameters	**N**	**Infants without Refeeding Syndrome (n = 471)**	**Infants with Refeeding Syndrome (n = 289)**	***p*-Value**	**N**	**Infants without Refeeding Syndrome (n = 128)**	**Infants with Refeeding Syndrome (n = 136)**	***p*-Value**
Average parenteral carbohydrate intake within the first 7 days (mg/kg/min), median (IQR)	760	8.47 (7.74–9.12)	8.46 (7.65–9.04)	0.55	264	8 (7–8.8)	7.7 (6.4–8.5)	0.03 *
Average parenteral phosphate intake within the first 7 days (mg/kg/min), median (IQR)	760	0.30 (0–0.6)	0.24 (0–0.4)	0.01 *	264	0.30 (0.02–0.62)	0.2 (0–0.44)	0.03 *
TPN duration, median (IQR)	760	14 (7–29)	18 (9–35)	0.002 *	264	28 (12–48)	23 (11–41)	0.45

* *p*-values < 0.05.

Infants with RFS had a lower gestational age and birth weight, shorter length, and smaller head circumference than those without RFS ($p < 0.001$).

Infants with RFS had lower Apgar scores at 1 and 5 min than infants without RFS ($p < 0.001$). Moreover, male infants were at higher risk of RFS than female infants ($p < 0.001$). In addition, infants who developed RFS required more surfactant than those without RFS ($p < 0.001$). Moreover, there was a significant association between average lipid intake, amino acid intake, phosphate intake, TPN duration, and RFS. Infants with RFS received more lipid, more amino acid, less phosphate, and more TPN than those without RFS (p = 0.02, 0.01, 0.01, and 0.002; respectively). Figure 2 shows the percentage of infants with RFS and neonatal outcomes.

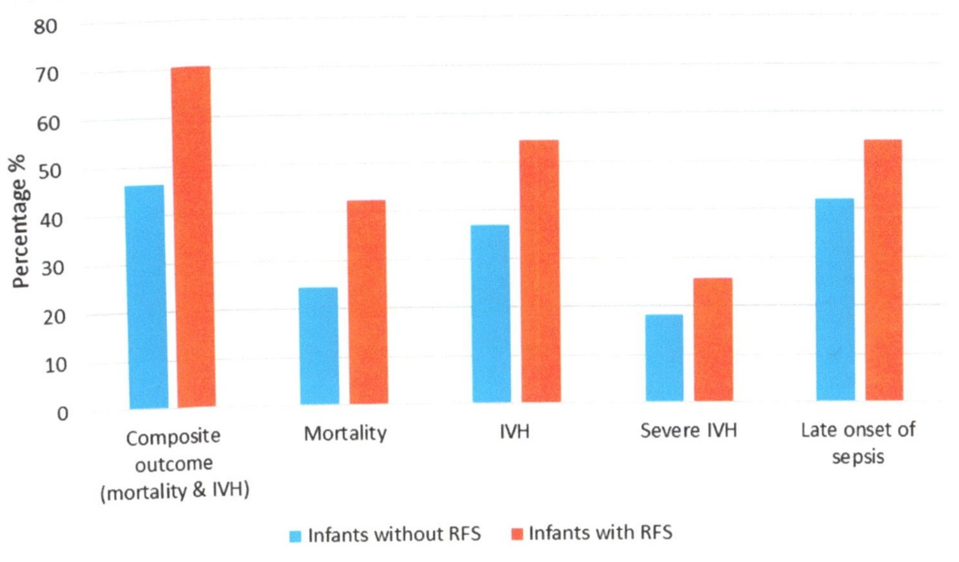

Figure 2. Percentage of neonatal morbidities and mortality stratified according to RFS.

In the univariate analysis, RFS was significantly associated with neonatal morbidity and mortality in very-preterm infants (Table 3). In infants aged < 32 gestational weeks, RFS was significantly associated with composite outcome, mortality, IVH, severe IVH, LOS, and BPD (p < 0.001, <0.001, <0.001, <0.001, <0.001, and 0.02, respectively) (Table 3). There was no significant difference between RFS and infants' growth anthropometrics. Moreover, no association was observed between RFS and the length of invasive ventilator as well as the length of hospital stay (p = 0.33, 0.05; respectively).

Table 3. Univariate analysis of growth anthropometrics, neonatal morbidity, and mortality stratified according to RFS.

Variables	<32 Weeks				<28 Weeks			
	N	Infants without Refeeding Syndrome (n = 471)	Infants with Refeeding Syndrome (n = 289)	p-Value	N	Infants without Refeeding Syndrome (n = 128)	Infants with Refeeding Syndrome (n = 136)	p-Value
Growth anthropometrics								
Weight gain velocity (g/kg/day), median (IQR)	574	6.7 (5.23–8.26)	6.78 (4.89–8.09)	0.50	141	6.16 (5.33–7.59)	6.61 (4.99–8.04)	0.72
Weight at discharge (g), median (IQR)	574	1850 (1770–2026)	1870 (1750–2020)	0.94	141	1935 (1790–2191)	1920 (1735–2347)	0.87
Z-score for weight at discharge, median (IQR)	574	−2.02 (−2.86 to −1.26)	−1.98 (−3.20 to −1.35)	0.41	141	−2.46 (−3.47 to −1.31)	−2.0 (−3.32 to −1.29)	0.40
Neonatal morbidities and mortality								
Composite outcome (IVH, mortality)	760	148 (31.4)	204 (70.5)	<0.001 *	264	73 (57)	120 (88.2)	<0.001 *
Mortality, n (%)	760	64 (13.6)	122 (42.2)	<0.001 *	264	39 (30.5)	84 (61.8)	<0.001 *
Intraventricular hemorrhage, n (%)	760	124 (26.3)	157 (54.3)	<0.001*	264	64 (50)	96 (70.6)	0.001 *
Severe intraventricular hemorrhage, n (%)	760	63 (13.3)	74 (25.6)	<0.001 *	264	41 (32)	56 (41.2)	0.15
Early-onset sepsis (culture-proven), n (%)	760	6 (1.3)	5 (1.7)	0.35	264	2 (1.6)	4 (2.9)	0.68
Late-onset sepsis (culture-proven), n (%)	760	164 (34.8)	156 (54)	<0.001 *	264	78 (60.9)	97 (71.3)	0.09
Necrotizing enterocolitis (stage ≥ 2), n (%):								
Medical management	760	147 (31.2)	95 (32.8)	0.63	264	61 (47.7)	46 (33.8)	0.02 *
Surgical management	760	30 (6.4)	19 (6.6)	1	264	18 (14.1)	15 (11)	0.46
Spontaneous intestinal perforation, n (%)	760	5 (1.1)	5 (1.7)	0.52	264	3 (2.3)	3 (2.2)	1
Bronchopulmonary dysplasia, n (%)	511	132 (40.4)	94 (51.1)	0.02 *	176	70 (72.9)	60 (75)	0.86
Duration of invasive ventilation, days, median (IQR)	760	11 (2–27)	13 (6–31)	0.3	264	13 (3–31)	17 (8–34)	0.29
Length of hospital stay after excluding death infants, median (IQR)	574	43 (26–67)	49 (31–73)	0.05	141	82 (62–102)	75 (55–105)	0.61

* p-values < 0.05.

Based on the multivariate regression analysis performed after adjusting the variables that were significant in the univariate analysis, RFS was significantly associated with composite outcome and IVH (adjusted relative risk [aRR]: 1.70, 95% confidence interval [CI]: 1.45–2.0; aRR: 1.59, 95% CI: 1.31–1.94, respectively) (Table 4).

Figure 3 shows the Kaplan–Meier curves of the association between RFS and survival among all infants (log-rank test, $p < 0.001$). The multivariate Cox regression curve adjusted for significant potential confounders showed a significant association between RFS and survival. Moreover, the hazard ratios (HRs) for death in infants with RFS increased by 1.74-fold (95% CI: 1.37–2.21) than those in infants without RFS (Table 4).

Table 4. Multivariate regression for infant's outcome stratified according to RFS.

Variables	ALL Neonates Unadjusted 95% CI	ALL Neonates Adjusted 95% CI	Newborns with Gestational Age < 28 Weeks Unadjusted 95% CI	Newborns with Gestational Age < 28 Weeks Adjusted 95% CI
Composite outcome (IVH, mortality) (RR)	2.25 (1.94–2.63)	1.70 (1.45–2.0)	1.55 (1.32–1.82)	1.40 (1.17–1.68)
Mortality (HR)	3.21 (2.37–4.34)	1.74 (1.37–2.21)	2.57 (1.75–3.77)	1.59 (1.20–2.09)
Intraventricular hemorrhage (RR)	2.07 (−9.08 to 13.22)	1.59 (1.31–1.94)	1.41 (1.15–1.73)	1.30 (1.06–1.65)
Severe intraventricular hemorrhage (RR)	1.93 (1.43–2.61)	1.24 (0.90–1.70)	1.28 (0.93–1.76)	----------
Bronchopulmonary dysplasia (RR)	1.27 (1.04–1.54)	0.86 (0.72–1.03)	1.03 (0.86–1.23)	----------
Late-onset sepsis (RR)	1.55 (1.32–1.83)	1.16 (0.98–1.39)	1.17 (0.98–1.3)	----------

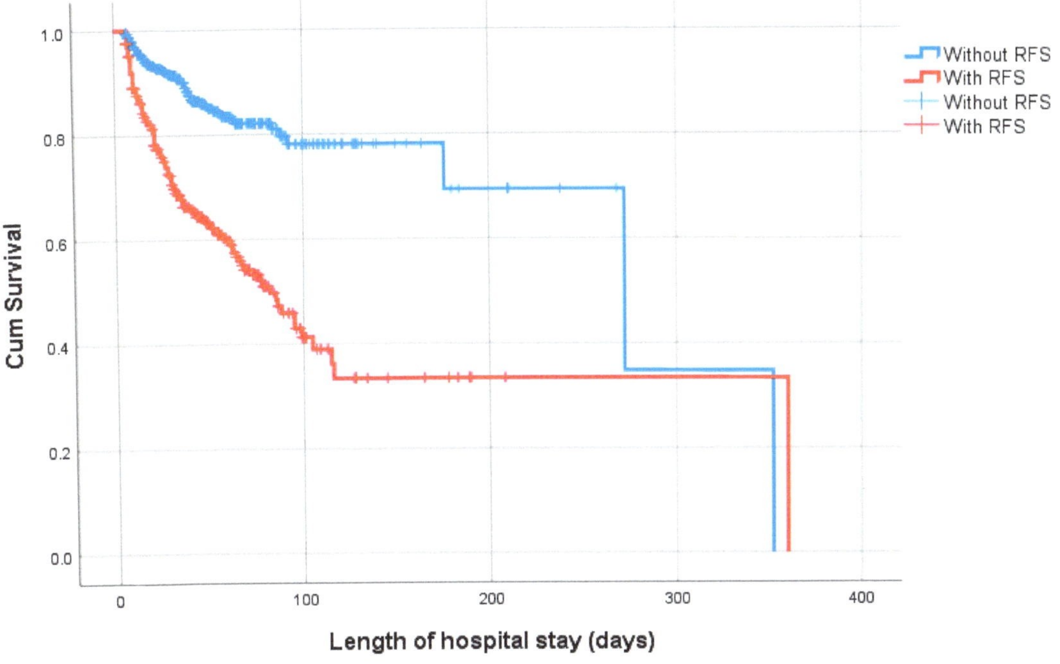

Figure 3. Kaplan–Meier curves showing the association between RFS and survival among infants aged < 32 gestational weeks, log-rank test, $p < 0.001$.

In the subanalysis, RFS was significantly associated with composite outcome, mortality, IVH, and NEC (medical) in newborns with gestational age < 28 gestational weeks ($p < 0.001$, $p < 0.001$, $p = 0.001$, and $p = 0.02$, respectively) (Table 3). The adjusted model of the multivariate regression analysis showed a significant association between RFS as well as composite outcome and IVH (aRR: 1.40, 95% confidence interval [CI]: 1.17–1.68); aRR: 1.30, 95% CI: 1.06–1.65, respectively) (Table 4).

Figure 4 shows the Kaplan–Meier curves of the association between RFS and survival among infants aged < 28 gestational weeks (log-rank test, $p < 0.001$). In newborns with gestational age < 28 gestational weeks, the multivariable Cox regression curve adjusted for significant potential confounders showed that the HRs for death in infants with RFS increased by 1.59-fold (95% CI: 1.20–2.09) compared with those in infants without RFS (Table 4).

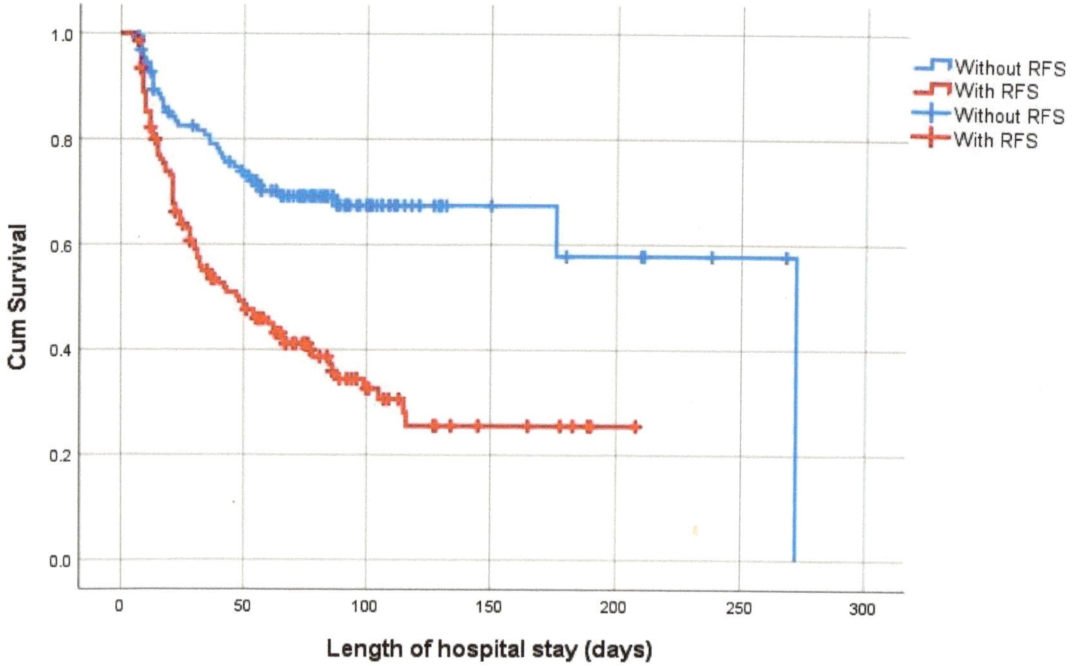

Figure 4. Kaplan–Meier curves showing the association between RFS and survival among infants aged < 28 gestational weeks, log-rank test, $p < 0.001$.

4. Discussion

The current study showed that preterm infants born at ≤32 weeks of gestation who developed RFS within the first week of life had a higher incidence of IVH and/or mortality during their hospital stay. However, there was no significant difference in the duration of mechanical ventilation, the rate of BPD, and the incidence of LOS between infants who developed RFS and those who did not.

Our findings are in accordance with those of the ProVIDe trial, which revealed that the incidence of mortality increased by 3-fold in infants with RFS and that severe hypophosphatemia was associated with a 5-fold increase in the rate of IVH [14]. In our study, the mortality rate of infants with RFS was approximately 42%. Meanwhile, despite using the same definition of RFS, the mortality rate in the ProVIDe trial was 32%. In addition, the mortality rate in the non-RFS population was similar between the two studies, with 13% in the current cohort versus 11% in the ProVIDe trial. In our previous study investigating the incidence and risk factors of RFS in preterm infants, IVH was found to be a significant risk factor for the development of RFS [16]. Considering that the majority of IVH cases occur within the first week of life, the temporal sequence between the onset of IVH and RFS is challenging to determine. Nonetheless, our findings showed a strong association between these two conditions in preterm infants. These results are in accordance with those of the ProVIDe trial, and the study conducted by Al-Wassia et al. also revealed a significant association between IVH and the development of RFS [14,17].

In contrast, some studies did not observe a significant difference in the incidence of mortality and IVH associated with RFS. These studies often had limitations. For example, they had small sample sizes, a low incidence of IVH, different definitions of RFS, and variations in the composition of TPN, which might have influenced their findings [9,19,21,22,24]. We did not find a significant difference in the rate of LOS between the two groups, a result consistent with the findings of other studies, including the ProVIDe trial [9,14,17,22,23,25]. However, Moltu et al. reported that VLBW infants with RFS had a high rate of LOS [24].

There was no convincing evidence showing that RFS is associated with a longer duration of mechanical ventilation or a higher rate of BPD. Our findings on the length of mechanical ventilation and BPD are in accordance with those of Ross et al. and Igarashi et al. [9,21] as well as the ProVIDe trial and study of Bustos et al., respectively [14,23]. However, other studies, such as those by Sung et al. and Ross et al., have reported associations between RFS and the duration of mechanical ventilation, and the incidence rate of BPD, respectively [9,22].

The discrepancies in the results of published papers could be attributed to different risk factors present in each population. In our study, male sex, IVH, and low sodium phosphate intake within the first week were significant risk factors for developing RFS in preterm infants [16]. Other studies have revealed that IUGR and SGA status are factors associated with an increased risk and severity of RFS [9,20,26,27]. In addition, some studies have found that the incidence of RFS has increased after the implementation of recommendations to increase amino acid intake within the first days of life [14,24,33].

Although there are differences among various studies, including our own, it is more important to recognize the risks associated with RFS in preterm infants and its potential short- and long-term complications. To address RFS effectively, it is important to establish a unified RFS definition in the NICU, which could help eliminate variations worldwide. In addition, clinical trials should be performed to determine the optimal balance of PN. Meanwhile, clinicians should implement a standardized approach in the NICU, involving a multidisciplinary team to facilitate the following: (1) identify at-risk infants; (2) develop individualized PN strategies; (3) establish guidelines for early enteral nutrition; and (4) create a laboratory monitoring protocol. These steps are essential, considering our findings, which show a significant association between RFS and adverse outcomes such as IVH and an increased rate of mortality.

Our current study was based on previous research on the incidence and risk factors of RFS in preterm infants, with a focus on highlighting the short-term complications associated with RFS in this population. This was a retrospective study conducted in a single tertiary center. However, it included a substantial number of patients who are at risk of developing RFS. In addition, infants with IUGR in this study were not identified, as this requires mothers to be closely followed throughout early antenatal care.

5. Conclusions

Preterm infants born at ≤32 weeks of gestation who develop RFS within the first week of life are at higher risk of IVH, mortality, and longer duration of TPN. These findings emphasize the urgent need for the early identification and standardized management of RFS in the NICU.

Author Contributions: Conceptualization, M.M.A.-M. and S.S.A.; data curation, T.M.K., R.S.A., A.O.A., Y.A.A. and M.R.A.-A.; formal analysis, S.S.A.; methodology, M.M.A.-M., B.A., A.A. and T.E.Y.; writing—original draft, M.M.A.-M., H.H.S., A.A.A., N.A.A. and S.S.A.; created the figures and tables, A.S.A., L.A. and S.S.A.; writing—review and editing, M.M.A.-M., B.A., H.H.S., A.A.A., N.A.A., L.A., A.A., T.E.Y., A.S.A., Y.A.A., T.M.K., R.S.A., A.O.A., M.R.A.-A. and S.S.A. All authors have read and agreed to the published version of the manuscript.

Funding: This research received no external funding.

Institutional Review Board Statement: This study was conducted in accordance with the Declaration of Helsinki and Good Pharmacoepidemiology Practice Guidelines and was approved by the Medical Ethical Review Committee of KSMC (reference number H1RI-12-May24-01, approved on 12 May 2024).

Informed Consent Statement: The requirement for consent was waived.

Data Availability Statement: The original contributions presented in the study are included in the article, further inquiries can be directed to the corresponding author.

Acknowledgments: We are particularly grateful to all the people who have given us help on our article.

Conflicts of Interest: The authors declare no conflicts of interest.

References

1. da Silva, J.S.V.; Seres, D.S.; Sabino, K.; Adams, S.C.; Berdahl, G.J.; Citty, S.W.; Cober, M.P.; Evans, D.C.; Greaves, J.R.; Gura, K.M.; et al. Parenteral Nutrition Safety and Clinical Practice Committees, American Society for Parenteral and Enteral Nutrition. ASPEN consensus recommendations for refeeding syndrome. *Nutr. Clin. Pract.* **2020**, *35*, 178–195. [CrossRef] [PubMed]
2. Schnitker, M.A.; Mattman, P.E.; Bliss, T.L. A clinical study of malnutrition in Japanese prisoners of war. *Ann. Intern. Med.* **1951**, *35*, 69–96.
3. Finn, P.F.; Dice, J.F. Proteolytic and lipolytic responses to starvation. *Nutrition* **2006**, *22*, 830–844. [CrossRef] [PubMed]
4. Cahill, G.F., Jr. Fuel metabolism in starvation. *Annu. Rev. Nutr.* **2006**, *26*, 1–22. [CrossRef]
5. Friedli, N.; Stanga, Z.; Sobotka, L.; Culkin, A.; Kondrup, J.; Laviano, A.; Mueller, B.; Schuetz, P. Revisiting the refeeding syndrome: Results of a systematic review. *Nutrition* **2017**, *35*, 151–160. [CrossRef] [PubMed]
6. Goretti Penido, M.; Alon, U.S. Phosphate homeostasis and its role in bone health. *Pediatr. Nephrol.* **2012**, *27*, 2039–2048. [CrossRef]
7. Sharma, S.; Brugnara, C.; Betensky, R.A.; Waikar, S.S. Reductions in red blood cell 2,3-diphosphoglycerate concentration during continuous renal replacment therapy. *Clin. J. Am. Soc. Nephrol.* **2015**, *10*, 74–79. [CrossRef]
8. Kovacs, C.S. Maternal mineral and bone metabolism during pregnancy, lactation, and post-weaning recovery. *Physiol. Rev.* **2016**, *96*, 449–547. [CrossRef]
9. Ross, J.R.; Finch, C.; Ebeling, M.; Taylor, S.N. Refeeding syndrome in very-low-birth-weight intrauterine growth-restricted neonates. *J. Perinatol.* **2013**, *33*, 717–720. [CrossRef]
10. Amissah, E.A.; Brown, J.; Harding, J.E. Protein supplementation of human milk for promoting growth in preterm infants. *Cochrane Database Syst. Rev.* **2018**, *6*, CD000433. [CrossRef]
11. Bellagamba, M.P.; Carmenati, E.; D'Ascenzo, R.; Malatesta, M.; Spagnoli, C.; Biagetti, C.; Burattini, I.; Carnielli, V.P. One extra gram of protein to preterm infants from birth to 1800 g: A single-blinded randomized clinical trial. *J. Pediatr. Gastroenterol. Nutr.* **2016**, *62*, 879–884. [CrossRef] [PubMed]
12. Agostoni, C.; Buonocore, G.; Carnielli, V.P.; De Curtis, M.; Darmaun, D.; Decsi, T.; Domellöf, M.; Embleton, N.D.; Fusch, C.; Genzel-Boroviczeny, O.; et al. Enteral nutrient supply for preterm infants: Commentary from the European Society of Paediatric Gastroenterology, Hepatology and Nutrition Committee on Nutrition. *J. Pediatr. Gastroenterol. Nutr.* **2010**, *50*, 85–91. [CrossRef] [PubMed]
13. Embleton, N.D.; Jennifer Moltu, S.; Lapillonne, A.; van den Akker, C.H.P.; Carnielli, V.; Fusch, C.; Gerasimidis, K.; van Goudoever, J.B.; Haiden, N.; Iacobelli, S.; et al. Enteral nutrition in preterm infants (2022): A position paper from the ESPGHAN Committee on Nutrition and invited experts. *J. Pediatr. Gastroenterol. Nutr.* **2023**, *76*, 248–268. [CrossRef] [PubMed]
14. Cormack, B.E.; Jiang, Y.; Harding, J.E.; Crowther, C.A.; Bloomfield, F.H.; ProVIDe Trial Group. Neonatal refeeding syndrome and clinical outcome in extremely low-birth-weight babies: Secondary cohort analysis from the ProVIDe trial. *JPEN J. Parenter. Enter. Nutr.* **2021**, *45*, 65–78. [CrossRef] [PubMed]
15. Weintraub, A.S.; Blanco, V.; Barnes, M.; Green, R.S. Impact of renal function and protein intake on blood urea nitrogen in preterm infants in the first 3 weeks of life. *J. Perinatol.* **2015**, *35*, 52–56. [CrossRef]
16. Asfour, S.S.; Alshaikh, B.; Mathew, M.; Fouda, D.I.; Al-Mouqdad, M.M. Incidence and risk factors of refeeding syndrome in preterm infants. *Nutrients* **2024**, *16*, 2557. [CrossRef]
17. Al-Wassia, H.; Lyon, A.W.; Rose, S.M.; Sauve, R.S.; Fenton, T.R. Hypophosphatemia is prevalent among preterm infants less than 1,500 grams. *Am. J. Perinatol.* **2019**, *36*, 1412–1419. [CrossRef]
18. Brener Dik, P.H.; Galletti, M.F.; Bacigalupo, L.T.; Fernández Jonusas, S.; Mariani, G.L. Hypercalcemia and hypophosphatemia among preterm infants receiving aggressive parenteral nutrition. *Arch. Argent. Pediatr.* **2018**, *116*, e371–e377.
19. Pająk, A.; Królak-Olejnik, B.; Szafrańska, A. Early hypophosphatemia in very low birth weight preterm infants. *Adv. Clin. Exp. Med.* **2018**, *27*, 841–847. [CrossRef]
20. Chaudhary, V.; Jayendra, R.G.; Shreya, A.P. Hypophosphatemia in refeeding syndrome in intrauterine growth restricted IUGR neonates who are receiving nutrition: A prospective observational study. *Asian J. Pediatr. Res.* **2020**, *3*, 23–29. [CrossRef]
21. Igarashi, A.; Okuno, T.; Ohta, G.; Tokuriki, S.; Ohshima, Y. Risk factors for the development of refeeding syndrome-like hypophosphatemia in very low birth weight infants. *Dis. Markers* **2017**, *2017*, 9748031. [CrossRef] [PubMed]
22. Sung, S.I.; Chang, Y.S.; Choi, J.H.; Ho, Y.; Kim, J.; Ahn, S.Y.; Park, W.S. Increased risk of refeeding syndrome-like hypophosphatemia with high initial amino acid intake in small-for-gestational-age, extremely-low-birthweight infants. *PLoS ONE* **2019**, *14*, e0221042. [CrossRef] [PubMed]
23. Bustos Lozano, G.; Soriano-Ramos, M.; Pinilla Martín, M.T.; Chumillas Calzada, S.; García Soria, C.E.; Pallás-Alonso, C.R. Early hypophosphatemia in high-risk preterm infants: Efficacy and safety of sodium glycerophosphate from first day on parenteral nutrition. *JPEN J. Parenter. Enteral Nutr.* **2019**, *43*, 419–425. [CrossRef] [PubMed]

24. Moltu, S.J.; Strømmen, K.; Blakstad, E.W.; Almaas, A.N.; Westerberg, A.C.; Brække, K.; Rønnestad, A.; Nakstad, B.; Berg, J.P.; Veierød, M.B.; et al. Enhanced feeding in very-low-birth-weight infants may cause electrolyte disturbances and septicemia—A randomized, controlled trial. *Clin. Nutr.* **2013**, *32*, 207–212. [CrossRef]
25. Brener Dik, P.H.; Galletti, M.F.; Fernández Jonusas, S.A.; Alonso, G.; Mariani, G.L.; Fustiñana, C.A. Early hypophosphatemia in preterm infants receiving aggressive parenteral nutrition. *J. Perinatol.* **2015**, *35*, 712–715. [CrossRef] [PubMed]
26. Boubred, F.; Herlenius, E.; Bartocci, M.; Jonsson, B.; Vanpée, M. Extremely preterm infants who are small for gestational age have a high risk of early hypophosphatemia and hypokalemia. *Acta Paediatr.* **2015**, *104*, 1077–1083. [CrossRef]
27. Ichikawa, G.; Watabe, Y.; Suzumura, H.; Sairenchi, T.; Muto, T.; Arisaka, O. Hypophosphatemia in small for gestational age extremely low birth weight infants receiving parenteral nutrition in the first week after birth. *J. Pediatr. Endocrinol. Metab.* **2012**, *25*, 317–321. [CrossRef] [PubMed]
28. Senterre, T.; Abu Zahirah, I.; Pieltain, C.; de Halleux, V.; Rigo, J. Electrolyte and mineral homeostasis after optimizing early macronutrient intakes in VLBW infants on parenteral nutrition. *J. Pediatr. Gastroenterol. Nutr.* **2015**, *61*, 491–498. [CrossRef] [PubMed]
29. Mbethe, A.P.; Mda, S. Incidence of refeeding syndrome and its associated factors in South African children hospitalized with severe acute malnutrition. *Iran. J. Pediatr.* **2017**, *27*, e8297. [CrossRef]
30. Papile, L.A.; Burstein, J.; Burstein, R.; Koffler, H. Incidence and evolution of subependymal and intraventricular hemorrhage: A study of infants with birth weights less than 1,500 gm. *J. Pediatr.* **1978**, *92*, 529–534. [CrossRef]
31. Ballabh, P. Intraventricular hemorrhage in premature infants: Mechanism of disease. *Pediatr. Res.* **2010**, *67*, 1–8. [CrossRef] [PubMed]
32. Huvanandana, J.; Nguyen, C.; Thamrin, C.; Tracy, M.; Hinder, M.; McEwan, A.L. Prediction of intraventricular haemorrhage in preterm infants using time series analysis of blood pressure and respiratory signals. *Sci. Rep.* **2017**, *7*, 46538. [CrossRef] [PubMed]
33. Bonsante, F.; Iacobelli, S.; Latorre, G.; Rigo, J.; De Felice, C.; Robillard, P.Y.; Gouyon, J.B. Initial amino acid intake influences phosphorus and calcium homeostasis in preterm infants—It is time to change the composition of the early parenteral nutrition. *PLoS ONE* **2013**, *8*, e72880. [CrossRef] [PubMed]

nutrients

MDPI

Article

Adherence to Nutritional Practice Guideline in Premature Infants: A Nationwide Survey in Taiwan

Chi-Shiuan Ting [1,2], Po-Nien Tsao [1], Hung-Chieh Chou [1], Ting-An Yen [1], Hsin-Chung Huang [1] and Chien-Yi Chen [1,*]

1 Department of Pediatrics, National Taiwan University Hospital, National Taiwan University College of Medicine, Taipei 100, Taiwan; chishiuan0212@gmail.com (C.-S.T.); tsaopn@ntu.edu.tw (P.-N.T.); hcchou@ntu.edu.tw (H.-C.C.); anneofgb@gmail.com (T.-A.Y.); newborn.huang@gmail.com (H.-C.H.)

2 Department of Pediatrics, China Medical University Children's Hospital, China Medical University, Taichung 404, Taiwan

* Correspondence: chienyichen@ntu.edu.tw; Tel.: +886-2-23123456-271015

Abstract: Objectives: This study aimed to assess the current neonatal nutritional practices in Taiwan and promote consensus on standardized protocols. Methods: An online questionnaire comprising 95 items on parenteral nutrition (PN) and enteral nutrition (EN) practices was distributed to neonatal care units across Taiwan via email between August and December 2022. The responses were compared with the recommendations from the European Society for Pediatric Gastroenterology Hepatology and Nutrition for preterm infant care. Results: Most of the 35 neonatal units, comprising 17 level III and 18 level II units, that participated in this study adhered to standard PN protocols; however, only 30% of units used protein-containing solutions as the initial fluid. Over half of the neonatal units provided calcium, phosphate, and magnesium at less than the recommended dosage. Trophic feeding commenced within 48 h in 88% of the units, with the mother's milk used as the first choice. All the units preferred commencing advanced feeding at <25 mL/kg/day. Conclusions: Most nutrient protocols for preterm infants in neonatal units in Taiwan meet recent guidelines, but discrepancies such as lower mineral supplements in PN and a slower advancement of enteral feeding increase nutritional risk. These issues warrant further research.

Keywords: neonatology; prematurity; parenteral nutrition; enteral nutrition; clinical practice

1. Introduction

Inadequate nutritional support results in suboptimal growth and poor developmental outcomes in preterm neonates [1–4]. Therefore, it is important to implement clinical practice guidelines that sufficiently address the nutritional needs of preterm infants. The consensus continues to evolve with accumulating evidence. Early aggressive parenteral nutrition (PN) with protein and lipid supplementation has been recommended in recent years [3]. An adequate intake of protein and minerals via PN plays an important role in enabling optimal growth [5,6]. An early initiation of enteral feeding and increasing the intake of breast milk, including donor human milk, support gut maturation and prevent complications in preterm infants. The most recent guidelines on neonatal nutrition, encompassing both parenteral and enteral methods, were developed via the regular revision of nutritional guidelines over decades.

The European Society of Pediatric Gastroenterology, Hepatology, and Nutrition (ESPGHAN) revised the guidelines for PN of preterm infants in 2018, which are considered as the most credible reference [5–13]. They also issued a position paper on enteral nutrition (EN) for preterm infants in 2022 [14]. Clinical practice often varies by region and context despite these guidelines [15,16]. The Taiwan Society of Neonatology published "Recommendation on nutritional care of preterm infants in Taiwan" in 2015 for the detailed PN and

EN policy, and it was revised in 2022, which follows the ESPGHAN recommendation [17]. However, clinical practices vary across institutions due to different reasons.

This study aimed to evaluate the implementation of the policies regarding PN and EN for preterm infants with a very low birth weight (VLBW) in neonatal care units in Taiwan and compare these implementations with the ESPGHAN guideline. We aim to conduct a survey to assess the current neonatal nutritional practices across Taiwan and promote consensus on standardized protocols.

2. Materials and Methods

A nationwide survey was conducted in all hospitals registered in the Taiwan Neonatal Network and the Premature Baby Foundation of Taiwan, which comprise 52 neonatal care units involved in providing over 90% of care for prematurity. This study was approved by the Ethics Committee of the National Taiwan University Hospital.

The nutritional survey was conducted using a closed-response online questionnaire designed in accordance with the ESPGHAN guidelines and clinical practices. Research Electronic Data Capture (RedCap) platform was used to conduct the survey. The questionnaire comprised 95 multiple-choice and open-ended questions subdivided into three domains. Domain I corresponded to the capacity of the neonatal units and comprised items pertaining to the number of beds in the NICU and observation room and the annual number of neonate admissions (including preterm and infants with VLBW). Domain II corresponded to PN and comprised items pertaining to the age at the time of initiation; initial dosage; and target intakes of carbohydrates, protein, lipids, minerals, and electrolytes. Domain III corresponded to the EN policy and comprised items pertaining to the priming schedule, rate of advanced feeding, and choice of feeding. In addition to these domains, items pertaining to the monitoring of metabolic bone disease in prematurity (MBDP) were included in this survey. The questionnaire was disseminated individually to the chief of each neonatal unit via electronic mail, and the responses based on the consensus and policy of each unit and the data were collected using REDCap 14.6.4 software. The data were excluded if a connection with the unit could not be achieved or no response was obtained by January 2023.

All the neonatal care units included in this study were classified as level II and level III units (classified according to the policy statement of the American Academy of Pediatrics) [18]. The data obtained from level II and level III neonatal care units were compared. The initiation and dosage of nutrients were analyzed and compared with the ESPGHAN guidelines on pediatric parenteral nutrition [5–12] and the World Health Organization (WHO) recommendations for the management of preterm or low-birth-weight infants [19] to determine compliance with local and global standards.

Statistical Analysis

All statistical analyses were performed using Statistical Analysis Software 9.4 M7 and Microsoft Excel 21 statistics. Continuous variables, expressed as median, range, and interquartile range (IQR), were compared using the Wilcoxon rank sum test. Comparisons between the categorical variables of the groups were performed using Fisher's exact test and Chi-square test owing to the small sample size. A p-value of <0.05 was considered statistically significant.

3. Results

A total of 35 of the 52 hospitals, including 17 of the 19 level III units (89.4%) and 18 of the 33 level II units (54.5%), responded to the survey before January 2023, yielding a response rate of 67.3%. Table 1 provides the annual number of preterm and VLBW infants admitted to these 35 hospitals. Approximately 20–25% of the infants born in Taiwan were delivered in these hospitals [20], accounting for 40% of preterm infants and 75% of VLBW infants.

Table 1. Demographic characteristics of respond institutions.

	Level III Units	Level II Units	Total
Number of units included in the study			
Units responded to questionnaire	17	18	35
Bed number of neonatal intensive care unit	15 (8–50)	6 (0–16)	435
Bed number of observation room	25 (12–54)	12 (5–22)	746
Number of neonatologists	5 (1–15)	2 (1–8)	133
Number of neonates per year			
Annual numbers of neonates in Taiwan			163,484
Admissions number of neonates	1300 (550–3168)	675 (367–1400)	38,694 (23.66%)
Annual numbers of prematurity in Taiwan			16,990
Admissions number of prematurity	230 (55–800)	72.5 (20–200)	6697 (39.4%)
Annual numbers of VLBW in Taiwan			1644
Admissions number of VLBW	45 (4–222)	7.5 (1–50)	1248 (75.9%)

This table is represented as the median number (Range) or number (percentage).

3.1. Parenteral Nutrition

Table 2 describes the results of the PN policy survey including fluid, carbohydrate, protein, lipid, and mineral intake.

Table 2. Parenteral nutrition policy for preterm infants.

	Total	Level III Units	Level II Units	*p*
Initial fluid intake				
70–90 mL/kg/day	32 (91%)	15 (88%)	17 (94%)	0.735
>90 mL/kg/day	1 (3%)	1 (6%)	0 (0%)	
Other	2 (6%)	1 (6%)	1 (6%)	
Target fluid intake				
<140 mL/kg/day	12 (34%)	6 (35%)	6 (33%)	1.00
>140 mL/kg/day	23 (66%)	11 (35%)	12 (67%)	
Initial fluid choice				
10% glucose	25 (71%)	11 (65%)	14 (78%)	0.47
Parenteral nutrition	10 (29%)	6 (35%)	4 (22%)	
Initial glucose intake				
<4 mg/kg/min	1 (3%)	1 (6%)	0 (0%)	0.485
4–8 mg/kg/min	34 (97%)	16 (94%)	18 (100%)	
Target glucose intake				
<8 mg/kg/min	14 (40%)	5 (29%)	9 (50%)	0.458
8–10 mg/kg/min	15 (43%)	8 (47%)	7 (39%)	
>10 mg/kg/min	4 (11%)	2 (12%)	2 (11%)	
Clinically	2 (6%)	2 (12%)	0 (0%)	
Age to initiated protein intake				
<24 h	22 (63%)	13 (76%)	9 (50%)	0.376
0–48 h	10 (28%)	3 (18%)	7 (39%)	
24–48 h	3 (9%)	1 (6%)	2 (11%)	

Table 2. *Cont.*

	Total		Level III Units		Level II Units		*p*
Target protein intake for birth weight ranging from 1001 to 1500 g							
<2.5 g/kg/day	1	(3%)	1	(6%)	0	(0%)	0.733
2.5–3.5 g/kg/day	20	(57%)	10	(59%)	10	(56%)	
>3.5 g/kg/day	14	(40%)	6	(35%)	8	(44%)	
Target protein intake for birth weight of <1000 g							
<2.5 g/kg/day	0	(0%)	0	(0%)	0	(0%)	1.00
2.5–3.5 g/kg/day	7	(20%)	3	(18%)	4	(24%)	
>3.5 g/kg/day	27	(80%)	14	(82%)	13	(76%)	
Age of initial lipid intake							
<24 h	4	(11%)	2	(12%)	2	(11%)	0.927
24–48 h	22	(63%)	12	(70%)	10	(56%)	
48–72 h	5	(14%)	2	(12%)	3	(17%)	
Clinically	2	(6%)	1	(6%)	1	(6%)	
Not given	2	(6%)	0	(0%)	2	(11%)	
Target lipid intake for birth weight ranging from 1001–1500 g							
2.0–2.4 g/kg/day	4	(12%)	3	(18%)	1	(6%)	0.804
2.5–2.9 g/kg/day	11	(33%)	6	(35%)	5	(31%)	
3.0–3.4 g/kg/day	16	(48%)	7	(41%)	9	(56%)	
3.5–3.9 g/kg/day	2	(7%)	1	(6%)	1	(6%)	
Target lipid intake for birth weight of <1000 g							
2.0–2.4 g/kg/day	3	(9%)	2	(12%)	1	(7%)	1.00
2.5–2.9 g/kg/day	9	(28%)	5	(29%)	4	(27%)	
3.0–3.4 g/kg/day	17	(53%)	9	(53%)	8	(53%)	
3.5–3.9 g/kg/day	3	(9%)	1	(6%)	2	(13%)	
Target Calcium intake							
20–39 mg/kg/day	2	(9%)	0	(0%)	2	(11%)	0.399
40–59 mg/kg/day	11	(28%)	4	(24%)	7	(39%)	
60–79 mg/kg/day	15	(53%)	9	(53%)	6	(33%)	
>80 mg/kg/day	3	(9%)	1	(6%)	2	(11%)	
Clinically	4	(11%)	3	(17%)	1	(6%)	
Organic Calcium used	27	(77%)	15	(88%)	12	(67%)	0.22
Target Phosphate intake							
<19 mg/kg/day	2	(6%)	0	(0%)	2	(11%)	<0.05
20–39 mg/kg/day	10	(29%)	3	(18%)	7	(39%)	
40–59 mg/kg/day	17	(49%)	11	(64%)	6	(33%)	
Clinically	3	(8%)	3	(18%)	0	(0%)	
Not given	3	(8%)	0	(0%)	3	(17%)	
Organic Phosphate used	14	(40%)	10	(59%)	4	(22%)	<0.05
Target Magnesium intake							
<5 mg/kg/day	21	(60%)	11	(65%)	10	(56%)	0.084
<5 mg/kg/day	8	(23%)	4	(24%)	4	(22%)	
Clinically	6	(17%)	2	(12%)	4	(22%)	

3.1.1. Fluids

Approximately 90% of the neonatal units initiated parenteral fluid intake at a dose of 70–90 mL/kg/day. The targeted fluid intake upon achieving stable growth was >140 mL/kg/day in nearly two-thirds of neonatal units. Most units used 10% dextrose solution without protein on the first day of life owing to the immediate unavailability of PN fluid in their units. This was particularly common in level II units, with 80% of the units using 10% dextrose solution.

3.1.2. Carbohydrates

The dosage of glucose administered varied widely across different institutes. The target dose of glucose was <8 mg/kg/min in half of the level II units and 29% of level III units. More than 90% of the surveyed units reduced the glucose infusion rate (GIR) when the blood glucose levels exceeded 250 mg/dL, rather than 180 mg/dL (as recommended by the recent guideline) [12]. The lowest GIR used during hyperglycemia varied across the units, ranging from 1.0 to 4.0 mg/kg/min. The glucose infusion was suspended in one unit if persistent hyperglycemia was observed. Most neonatologists in Taiwan tended to avoid the use of insulin in preterm infants.

3.1.3. Proteins

The current guideline recommends administering protein supplements in PN on the first postnatal day [3]. However, protein supplementation was commenced within the first postnatal day in only half of the surveyed units. The proportion of level III units (73%) was significantly higher than that of the level II units (50%). The target protein intake in infants with VLBW varied across units, with most units administering doses higher than those recommended by the current guideline. Approximately 40% of the units administered protein at a dose of >3.5 g/kg/day in preterm infants weighing 1000–1500 g at birth. Approximately 80% of the units reported a target protein intake of >3.5 g/kg/day for preterm infants with an extremely low birth weight (ELBW); no significant differences were observed between levels of care.

3.1.4. Lipids

Nearly 60% of the neonatal units initiated the administration of intravenous lipid emulsions on the second day of life. However, two level II units did not administer lipids to preterm neonates. The target dose for lipids, ranging from 2.0 to 3.9 g/kg/day, varied across units. Approximately half of the units aimed for a target lipid intake of 3.0–3.4 g/kg/day for infants with VLBW. Over 80% of the units administered SMOFlipid as lipid emulsions. However, two units still used a pure soybean oil-based lipid emulsion (INTRALIPID®).

3.1.5. Mineral Intake

The implementation of the calcium (Ca) supplementation policy across the units was consistent, with two-thirds of the neonatal units administering Ca at a dose of 40–79 mg/kg/day. However, compared with level II units, level III units tended to aim for a significantly higher target phosphate (P) intake, ranging from 40 to 59 mg/kg/day ($p < 0.05$). Nearly 60% of the neonatal units reported a magnesium (Mg) intake of <5 mg/kg/day in both groups. Recent guidelines from the ESPGHAN [6] recommend administering 64–140 mg/kg/day of Ca and 50–108 mg/kg/day of P in neonatal PN regimens to facilitate optimal growth and bone mineralization in preterm infants. Most units did not achieve the optimal target. Approximately 80% of the surveyed units used organic calcium as a supplement, with the prevalence of its use being higher in level III units. However, organic phosphate was used in only 40% of the neonatal care units, comprising approximately 60% of the level III units and 22% of the level II units. The proportion of level III units was significantly higher than that of the level II units ($p < 0.05$).

3.1.6. Monitoring

Some complications may arise during the use of parenteral nutrition, such as hypoglycemia or hyperglycemia, electrolyte imbalance, hyperlipidemia, and PN-associated liver disease. Therefore, the regular monitoring of associated parameters is suggested. Table 3 describes the monitoring schedule in preterm neonates using PN. Approximately 90% of the neonatal units regularly monitored nutrition-related biomarkers including the blood glucose, liver enzymes, electrolyte balance, and triglyceride levels. Electrolyte balance was the most frequently monitored biomarker, with 80% of the units monitoring it twice weekly or weekly. Approximately 70% of the units monitored the triglyceride levels weekly or biweekly, whereas nearly 70% of the departments monitored the liver enzyme levels on a weekly basis. Over 70% of the units did not monitor the coagulopathy factor levels regularly unless there were signs of bleeding.

Table 3. The monitor policy during use of parenteral nutrition in preterm infants.

	Once Daily		Twice a Week		Weekly		Biweekly		Monthly		Clinically		Total
Blood sugar	5	(16%)	8	(25%)	9	(28%)	0	(0%)	1	(3%)	9	(28%)	32
Electrolyte	1	(3%)	12	(38%)	15	(47%)	2	(6%)	1	(3%)	1	(3%)	32
Triglyceride	0	(0%)	0	(0%)	14	(44%)	9	(28%)	5	(16%)	4	(12%)	32
Liver function	0	(0%)	1	(3%)	23	(72%)	5	(16%)	3	(9%)	0	(0%)	32
Coagulation	0	(0%)	0	(0%)	2	(6%)	3	(10%)	3	(10%)	23	(74%)	31

3.2. Enteral Nutrition

Table 4 describes the survey results of enteral nutrition policy for infants with VLBW, including the choice of initial feeding, advancement of enteral feeding, and hospitalization for enteral nutrition for preterm infants.

Table 4. Enteral nutrition policy for preterm infants.

	Total		Level III Units		Level II Units		p
Start of trophic feeding							0.602
<48 h	31	(89%)	16	(94%)	15	(83%)	
48–96 h	4	(11%)	1	(6%)	3	(17%)	
Colostrum as mouth care							0.488
Yes	22	(63%)	12	(71%)	10	(56%)	
Choice of trophic feeding							0.16
MOM only	1	(3%)	1	(6%)	0	(0%)	
MOM > DM	21	(60%)	12	(70%)	9	(50%)	
MOM > PF	13	(37%)	4	(24%)	9	(50%)	
Definition of full feeding							0.102
<140 mL/kg/day	8	(23%)	4	(24%)	4	(22%)	
140–149 mL/kg/day	4	(11%)	4	(24%)	0	(0%)	
150–159 mL/kg/day	23	(66%)	9	(52%)	14	(78%)	
Rate of advance feeding							0.77
1–9 mL/kg/day	5	(14%)	3	(18%)	2	(11%)	
10–19 mL/kg/day	24	(69%)	12	(70%)	12	(67%)	
20–25 mL/kg/day	6	(17%)	2	(12%)	4	(22%)	
Time to add HMF							0.396
<60 mL/kg/day	2	(6%)	0	(0%)	2	(11%)	
60–80 mL/kg/day	4	(11%)	1	(6%)	3	(16%)	
80–100 mL/kg/day	15	(43%)	9	(53%)	6	(34%)	
100–120 mL/kg/day	13	(37%)	6	(35%)	7	(39%)	
Other	1	(3%)	1	(6%)	0	(0%)	
Time to stop PN							0.632
<100 mL/kg/day	2	(5%)	0	(0%)	2	(11%)	
100–120 mL/kg/day	17	(49%)	9	(53%)	8	(44%)	
120–140 mL/kg/day	14	(40%)	7	(41%)	7	(39%)	
>140 mL/kg/day	1	(3%)	0	(0%)	1	(6%)	
Other	1	(3%)	1	(6%)	0	(0%)	
Expected weight gain							0.470
<14 g/kg/day	2	(6%)	1	(6%)	1	(6%)	
15–24 g/kg/day	31	(88%)	14	(82%)	17	(94%)	
>25 g/kg/day	2	(6%)	2	(12%)	0	(0%)	

Table 4. *Cont.*

	Total		Level III Units		Level II Units		*p*
Regular vitamin D supply							
Yes	31	(89%)	16	(94%)	15	(83%)	0.602
Establish feeding protocol							
Yes	20	(57%)	12	(70%)	8	(44%)	0.175

3.2.1. Choice of Initial Feeding

Buccal colostrum was administered to premature infants initially in almost 60% of the neonatal units, with level III units (70%) constituting the majority. Approximately 90% of units initiated minimal enteral feeding within two days of birth, with mother's own milk (MOM) being the preferred choice in all units. However, donor human milk (DHM) was used as the second-line trophic feeding option in 70% of cases in level III units, compared to 50% in level II units. Additionally, half of the level II units chose 20 kcal/fl oz premature formula as their second-line option.

3.2.2. In-Hospital Enteral Nutrition

Approximately 70% of the surveyed units defined full feeding as ranging from 150 to 159 mL/kg/d for growing preterm infants. However, approximately 50% of level III units administered <150 mL/kg/d to infants who were completely dependent on EN. The advancement of enteral feed varied widely across the units, with 70% of the units selecting a slow advancement of 10–19 mL/kg/d. The 2022 WHO recommendation [19] states that the fast advancement (>30 mL/kg/day) of enteral feeding may reduce the time to full enteral feeding and the length of hospital stay. The majority of neonatal units tend to fortify the breast milk when the intake reaches 100 mL/kg/d. TPN was discontinued when the oral intake reached 120 mL/kg/d. Most neonatal units defined the standard of adequate weight gain as 15–24 g/kg/day after stable nutritional supplementation. More than 90% of the units implemented regular vitamin D supplementation for preterm infants. Thirty-two of the thirty-five surveyed units routinely monitored gastric residuals before every meal. However, 23 of these units continued feeding despite the presence of gastric residual volume. Over half of the units discontinued feeding if abdominal distension was suspected or an umbilical artery catheter was placed. Only 57% of the units, comprising 70% of level III units and 44% of level II units, implemented standardized feeding protocols.

3.2.3. Survey of Metabolic Bone Disease of Prematurity

As shown in Table 5, 28 units, including 16 level III units (94.1%) and 12 level II units (66.67%), routinely screened for MBDP before discharge. All units commenced MBDP screening not later than 5 weeks of birth, and 57% of the units performed the first biochemical test between 4 and 5 weeks of birth. The majority of the units utilized the serum calcium, phosphate, and alkaline phosphatase (ALP) levels as screening biomarkers. However, <40% of the units utilized additional biomarkers, such as 25-OH vitamin D, intact parathyroid hormone (iPTH), and urine biomarkers, for further screening. Four units performed bone mass measurements on premature infants.

Table 5. The monitor policy metabolic bone disease of prematurity.

Weekly		Biweekly		Monthly		Bimonthly		Not recorded		Clinically		Total
Alkaline Phosphatase												
0	(0%)	16	(57%)	11	(39%)	1	(3.5%)	0	(0%)	0	(0%)	28
Serum calcium/phosphate												
3	(11%)	16	(57%)	8	(29%)	1	(3.5%)	0	(0%)	0	(0%)	28

Table 5. *Cont.*

	Weekly	Biweekly	Monthly	Bimonthly	Not recorded	Clinically	Total
25-OH Vitamin D	0 (0%)	1 (4%)	5 (18.5%)	1 (4%)	15 (56%)	5 (18.5%)	27
Intact parathyroid hormone	0 (0%)	1 (3.5%)	5 (18%)	0 (0%)	17 (60%)	5 (18%)	28
Urine calcium/phosphate	0 (0%)	1 (3.5%)	4 (14%)	0 (0%)	19 (68%)	4 (14%)	28
Radiography	0 (0%)	6 (21%)	10 (36%)	3 (11%)	5 (18%)	4 (14%)	28
Bone Mass Measurement	0 (0%)	1 (3.5%)	1 (3.5%)	0 (0%)	24 (86%)	2 (7%)	28

4. Discussion

A comprehensive survey of the prevailing objectives pertaining to PN and EN for infants with VLBW in Taiwan was conducted in this study. Compared with the most recent guidelines for neonatal nutrition of infants with VLBW issued by the ESPGHAN [5–12], most objectives met the guidelines. However, certain items continued to differ between the international guidelines and the policies in Taiwan; these differences may be attributed to regional variations. In Taiwan, most neonatologists remain concerned about the association of feeding and the occurrence of NEC and tend to be more conservative in their feeding strategies. Limited medical resources is also one of the key factors affecting the compliance with recommendations. Although the National Health Insurance covers most of the high medical costs for preterm infants in NICU, the system imposes certain limitations on clinical care options. For example, some neonatal units cannot provide protein on the first day or higher calcium and phosphate intake by PN due to inadequate resources.

Over half of the surveyed units administered parenteral amino acid to preterm infants before the first day of life. Current guidelines recommend commencing the administration of amino acid intake in neonates from the first day of life to enhance protein synthesis while avoiding a decrease in proteolysis [21] as this approach can improve short-term growth [22,23]. However, the estimation of the required amino acid intake remains controversial. Almost 40% of the units surveyed in the current study administered >3.5 g/kg/day of parenteral amino acid to infants weighing 1000–1500 g. Over 80% of the units administered >3.5 g/kg/day of parenteral amino acid to infants with VLBW. The 2005 ESPGHAN guidelines recommend maintaining a minimum intake of 1.5 g/kg/day to prevent negative nitrogen balance, with a maximum of 4 g/kg/day [24]. Nevertheless, the evidence to support the contention that increasing the amino acid intake beyond 2.5 g/kg/day yields more favorable outcomes is limited. The effect of higher parenteral amino acid intake on long-term growth or neurodevelopment did not differ significantly [22,25,26]. Furthermore, high amino acid intake might elevate the urea concentration and sepsis rate. The 2018 ESPGHAN guidelines recommend maintaining the parenteral amino acid intake between 2.5 and 3.5 g/kg/day in preterm infants [5]. According to the recent guideline, the protein intake in this study was higher than what is currently recommended. This may be attributed to adhering to the previous 2005 ESPGHAN guideline. Nevertheless, the evidence for the influence of higher protein administration in preterm infants remains lacking.

The current study revealed that <50% of the units supplemented glucose infusion up to 8–10 mg/kg/min. In contrast, approximately 50% of the level II units aimed to maintain the infusion rate at <8 mg/kg/min. The optimal carbohydrate intake is determined based on the energy requirements, blood glucose levels, and growth. The glucose intake should be increased stepwise over 2–3 days, usually up to 10 mg/kg/min to facilitate growth thereafter. The ESPGHAN recommends that the parenteral carbohydrate intake should preferably not be >12 mg/kg/min or <4 mg/kg/min in preterm infants [12]. The blood glucose level is an important factor affecting the dose of glucose to be administered on the first postnatal day. Over 90% of the neonatal units adjusted the glucose infusion rate in response to blood glucose levels exceeding 250 mg/dL, with the minimum infusion rates varying across units. Hyperglycemia, defined as a blood glucose level of >180 mg/dL in preterm infants [27], is associated with increased morbidity [28–31]. However, the strict control of the blood glucose levels in critically ill children did not decrease mortality in previous studies [32,33]. Insulin therapy is effective in treating or preventing hyperglycemia in preterm infants, but it can lead to an increase in the incidence of hypoglycemia [34]. A general consensus on the management of blood sugar levels in premature infants has not been established in Taiwan, and the policy regarding the use of insulin in neonatal units remains unclear, warranting further research.

The target intake of Ca and Mg was similar across all neonatal care units in Taiwan in the current study. The target intake of phosphate was significantly higher in level III units. Ca, P, and Mg, which constitute 98%, 80%, and 65% of the body content, respectively, are major components of the skeleton. Optimal PN should provide a slight surplus of Ca and P to ensure optimal tissue and bone mineral accretion [6]. The fetal total body analysis suggested that the theoretical optimal molar Ca and P ratio in PN for achieving fetal body composition in stable growing infants is 1.3 [6]. Recent guidelines from the ESPGHAN recommend administering 64–140 mg/kg/day of Ca and 50–108 mg/kg/day of P as a part of neonatal PN regimens to facilitate optimal growth and bone mineralization in preterm infants [6]. Most neonatal units did not achieve the mineral intake levels recommended by the ESPGHAN recommended mineral intake. This may be attributed to the lower doses recommended by the previous guideline in Taiwan (Ca: 60–80 mg/kg/day, P: 45–60 mg/kg/day) [17,35]. This difference may be attributed to the previous absence of a published guideline for mineral intake in PN and the compatibility issues of inorganic salts in PN. Inadequate mineral intake is associated with MBDP, which may cause osteopenia and severe bone disease with fracture [36–38].

The use of organic phosphate (NaGP) in PN solutions can improve compatibility and prevent precipitation [39–41]. Only 40% of the surveyed neonatal units administered organic phosphate, possibly owing to the Taiwan Food and Drug Administration approving the use of NaGP in 2022. Although NaGP in PN solutions has been used to increase Ca and P intake in preterm infants receiving PN support, it does not facilitate adequate Ca and P intake in infants with ELBW [41]. Hsu et al. suggested that this result may be attributed to the high sodium content in NaGP, which limits the total P intake. An inappropriately high sodium intake may result in fluid retention and increase the risk of complications, such as hemodynamically significant patent ductus arteriosus, bronchopulmonary dysplasia (BPD), retinopathy of prematurity, and intraventricular hemorrhage, in preterm infants [41].

Most of the surveyed neonatal units implemented a similar strategy of preterm EN including time to start feeding, choice of trophic feeding, and rate of advance feeding. EN aims to support optimal growth and development of preterm infants while preventing the incidence of complications such as necrotizing enterocolitis (NEC), catheter-related complications, infections, and sepsis [42,43]. The ESPGHAN advises commencing enteral feeding at the earliest, preferably within the first few hours of life, and gradually increasing feedings to achieve full feeds by 2–3 weeks of age [14]. Human milk is the preferred feeding option for preterm infants, followed by donor milk or 24 kcal/fl oz premature formula if human milk is unavailable. Fortifying breast milk or premature formula with additional nutrients aids in meeting the elevated nutritional requirements of preterm infants. Approximately 90% of the surveyed units initiated minimal enteral feeding within two days of birth, and MOM was the first choice in all units.

Most of the surveyed units adopted similar feeding policies; however, certain differences were observed between the studies [14,19]. Several trials have indicated that fast advancement (>30 mL/kg/day) of enteral feeding may reduce the time to full enteral feeding and length of hospital stay without increasing the incidence of NEC [44,45]. However, all surveyed units preferred implementing a slow advancement policy for preterm infants. Over 50% of the units included in the current study preferred a rate of 10–19 mL/kg/day. This result could be attributed to the neonatologists in Taiwan being cautious about the risk of NEC and setting their own guidelines based on the findings of previous study [45]. The prevalence of NEC may be lower in Taiwan compared with that in other parts of the world owing to the implementation of the slow advancement policy. A global meta-analysis conducted in 2020 [46] showed that seven out of one hundred infants with VLBW (1.5–17%) were diagnosed with NEC and that the incidence rate of NEC from 2016 to 2021 was around 4.5% in Taiwan, which is slightly lower than the global incidence rate. Nevertheless, the correlation between fast advancement and the incidence of NEC remains unclear, warranting further studies. Rapid advancement in enteral feeding and the optimization of standard nutrition protocols are associated with potential benefits, such as increased body weight gain, decreased time to achieve full feeding, and reduced requirement for the prolonged use of PN [45,47]. These may help reduce the length of hospital stay and the incidence of PN-related complications, potentially benefiting neonatal neurodevelopment.

Twenty-eight of the thirty-five surveyed neonatal units screened for MBDP regularly; the screening commenced within 5 weeks. MBDP, a common complication observed in preterm infants recently, is characterized by suboptimal bone matrix mineralization and biochemical alterations of phospho-calcium metabolism. A lower birth weight, prolonged use of PN, and the incidence of NEC and BDP were identified as risk factors for MBDP [48]. The clinical signs of MBDP, which appear between 5 and 11 weeks of life, are characterized by an increased work of breathing (owing to chest wall instability caused by softening or fractures of ribs), an enlargement of the cranial sutures, frontal bossing, rickets, fractures, and postnatal growth failure [49,50]. The assessment of serum biochemical markers, such as Ca, P, ALP, iPTH, and vitamin D, can facilitate the early detection of mineral deficiency caused by MBDP [51–53]. The prevention strategy comprises improving nutrition, specifically the intake of Ca, P, and vitamin D, and limiting the chronic use of diuretics and methylxanthines that reduce mineral stores and glucocorticoids that enhance bone resorption. A biweekly monitoring of the above mentioned biochemical markers should be performed in infants at high risk of developing MBDP [48,54]. The long-term consequences of MBDP are difficult to study, partly owing to the difficulties in defining a diagnosis of MBDP and the lack of evidence regarding the correlation between MBDP and growth restriction or the incidence of fracture in childhood and adulthood [54]. Over 80% of the surveyed 28 units monitored the Ca, P, and ALP levels within a month; however, the vitamin D and iPTH levels were assessed less frequently. Previous studies have shown that the lack of mineral administration in PN, a slow advancement of feeding policy, and prolonged use of PN may increase the risk of developing MBDP. Diagnostic criteria for MBDP remains to be established; however, routine monitoring remains crucial among premature infants in Taiwan owing to the multitude of high-risk factors associated with our nutrition policy. However, the influence of MBDP and the complication warrant more evidence.

In summary, we have highlighted several key differences between international guidelines and current clinical practice in Taiwan. First, only 30% of units provide protein-containing solutions as the initial fluid. This is a systemic issue that requires more resources to improve. Secondly, over half of the neonatal units provided calcium, phosphate, and magnesium at less than the recommended dosage. The inadequate mineral intake in PN may contribute to premature osteopenia and increased fracture risk. Promoting the use of organic calcium and phosphate salts in PN and revising the existing guideline in Taiwan may improve the compliance. Finally, the practice of slow advancement of feeding is common in Taiwan, possibly due to the concerns of the risk of NEC. This feeding policy may lead to prolonged PN usage and extended hospital stays [45,47]. In our opinion, rather than applying a uniform slow feeding approach for all infants, feeding advancement should be customized based on clinical markers, such as gut function and tolerance.

Certain limitations exist to this study. First, the questionnaires were sent to the chiefs of the units, not the individual neonatologists. The variation between the nutritional objectives followed by each physician may have been higher than that observed between the nutritional protocols. However, the response from each unit represents the nutritional policy from the consensus formed among their medical staff. The result reflects the clinical practice of the majority of healthcare professionals. If the survey is collected on a per-person basis, the clinical practice of neonatal units with fewer doctors may be overlooked. Another limitation is that the charts of the infants were not reviewed to document actual feed. A comparison between the perceived and actual practices may reveal sizable variations [55]. There have been significant advancements in the clinical care of nutrition for preterm infants recently, but some practices have not yet reached a consensus or remain controversial. This questionnaire was developed in accordance with the ESPGHAN guidelines and focuses solely on the general principles of PN and EN care. Some issues of care that have not yet reached a consensus or remain controversial are not included in this survey. Some topics that have not yet reached a consensus in guideline were not included in this survey, such as the use of probiotics or in NICU.

5. Conclusions

In conclusion, the findings of the current study describe the PN and EN protocols for preterm infants in neonatal units in Taiwan. Most of these policies met the recent guidelines; however, certain discrepancies were observed between these practices and the recent policy. First, the target mineral intake of PN was lower than the national recommendation, especially in the level II units. Second, the advancement of enteral feeding policy tended to be slower (10–19 mL/kg/day) than what is recommended by the guidelines. These differences are risk factors for MBDP that warrant further research.

Author Contributions: Conceptualization, C.-S.T. and C.-Y.C.; methodology, C.-Y.C.; software, C.-Y.C.; validation, C.-Y.C.; formal analysis, C.-S.T.; data curation, C.-S.T., H.-C.C., T.-A.Y. and H.-C.H.; writing—original draft preparation, C.-S.T.; writing—review and editing, C.-Y.C. and P.-N.T.; project administration, C.-S.T. All authors have read and agreed to the published version of the manuscript.

Funding: This research received no external funding.

Institutional Review Board Statement: The study was conducted in accordance with the Declaration of Helsinki, and approved by the Research Ethics Committee, National Taiwan University Hospital (protocol code 202207152RINC and 12 August 2022).

Informed Consent Statement: Not applicable.

Data Availability Statement: The original contributions presented in the study are included in the article, further inquiries can be directed to the corresponding author.

Acknowledgments: The authors would like to thank all the physicians and units who generously shared their valuable experiences. Their professional knowledge and practical insights have been crucial to the successful completion of this research. Additionally, the authors are deeply grateful to the Premature Baby Foundation of Taiwan for providing relevant data and support.

Conflicts of Interest: The authors declare no conflicts of interest.

Abbreviations

The following abbreviations are used in this manuscript:

PN	Parenteral nutrition
EN	Enteral nutrition
ESPGHAN	European Society for Pediatric Gastroenterology Hepatology and Nutrition
TNN	Taiwan Neonatal Network
WHO	World Health Organization
VLBW	Very low birth weight
ELBW	extremely low birth weight
NICU	Neonatal intensive care unit

References

1. dit Trolli, S.E.; Kermorvant-Duchemin, E.; Huon, C.; Bremond-Gignac, D.; Lapillonne, A. Early lipid supply and neurological development at one year in very low birth weight (VLBW) preterm infants. *Early Hum. Dev.* **2012**, *88*, S25–S29. [CrossRef] [PubMed]
2. Stephens, B.E.; Walden, R.V.; Gargus, R.A.; Tucker, R.; McKinley, L.; Mance, M.; Nye, J.; Vohr, B.R. First-week protein and energy intakes are associated with 18-month developmental outcomes in extremely low birth weight infants. *Pediatrics* **2009**, *123*, 1337–1343. [CrossRef] [PubMed]
3. Eleni-dit-Trolli, S.; Kermorvant-Duchemin, E.; Huon, C.; Mokthari, M.; Husseini, K.; Brunet, M.L.; Dupont, C.; Lapillonne, A. Early individualised parenteral nutrition for preterm infants. *Arch. Dis. Child. Fetal Neonatal Ed.* **2009**, *94*, F152–F153. [CrossRef] [PubMed]
4. Ehrenkranz, R.A.; Dusick, A.M.; Vohr, B.R.; Wright, L.L.; Wrage, L.A.; Poole, W.K. Growth in the neonatal intensive care unit influences neurodevelopmental and growth outcomes of extremely low birth weight infants. *Pediatrics* **2006**, *117*, 1253–1261. [CrossRef]
5. van Goudoever, J.B.; Carnielli, V.; Darmaun, D.; Sainz de Pipaon, M.; Braegger, C.; Bronsky, J.; Cai, W.; Campoy, C.; Decsi, T.; Domellöf, M.; et al. ESPGHAN/ESPEN/ESPR/CSPEN guidelines on pediatric parenteral nutrition: Amino acids. *Clin. Nutr.* **2018**, *37*, 2315–2323. [CrossRef]
6. Mihatsch, W.; Fewtrell, M.; Goulet, O.; Molgaard, C.; Picaud, J.C.; Senterre, T.; Braegger, C.; Bronsky, J.; Cai, W.; Campoy, C.; et al. ESPGHAN/ESPEN/ESPR/CSPEN guidelines on pediatric parenteral nutrition: Calcium, phosphorus and magnesium. *Clin. Nutr.* **2018**, *37*, 2360–2365. [CrossRef]
7. Bronsky, J.; Campoy, C.; Braegger, C.; Braegger, C.; Bronsky, J.; Cai, W.; Campoy, C.; Carnielli, V.; Darmaun, D.; Decsi, T.; et al. ESPGHAN/ESPEN/ESPR/CSPEN guidelines on pediatric parenteral nutrition: Vitamins. *Clin. Nutr.* **2018**, *37*, 2366–2378. [CrossRef]
8. Hartman, C.; Shamir, R.; Simchowitz, V.; Lohner, S.; Cai, W.; Decsi, T.; Braegger, C.; Bronsky, J.; Campoy, C.; Carnielli, V.; et al. ESPGHAN/ESPEN/ESPR/CSPEN guidelines on pediatric parenteral nutrition: Complications. *Clin. Nutr.* **2018**, *37*, 2418–2429. [CrossRef]
9. Jochum, F.; Moltu, S.J.; Senterre, T.; Nomayo, A.; Goulet, O.; Iacobelli, S.; Braegger, C.; Bronsky, J.; Cai, W.; Campoy, C.; et al. ESPGHAN/ESPEN/ESPR/CSPEN guidelines on pediatric parenteral nutrition: Fluid and electrolytes. *Clin. Nutr.* **2018**, *37*, 2344–2353. [CrossRef]
10. Joosten, K.; Embleton, N.; Yan, W.; Senterre, T.; Braegger, C.; Bronsky, J.; Cai, W.; Campoy, C.; Carnielli, V.; Darmaun, D.; et al. ESPGHAN/ESPEN/ESPR/CSPEN guidelines on pediatric parenteral nutrition: Energy. *Clin. Nutr.* **2018**, *37*, 2309–2314. [CrossRef]
11. Lapillonne, A.; Fidler Mis, N.; Goulet, O.; van den Akker, C.H.P.; Wu, J.; Koletzko, B.; Braegger, C.; Bronsky, J.; Cai, W.; Campoy, C.; et al. ESPGHAN/ESPEN/ESPR/CSPEN guidelines on pediatric parenteral nutrition: Lipids. *Clin. Nutr.* **2018**, *37*, 2324–2336. [CrossRef] [PubMed]
12. Mesotten, D.; Joosten, K.; van Kempen, A.; Verbruggen, S.; Braegger, C.; Bronsky, J.; Cai, W.; Campoy, C.; Carnielli, V.; Darmaun, D.; et al. ESPGHAN/ESPEN/ESPR/CSPEN guidelines on pediatric parenteral nutrition: Carbohydrates. *Clin. Nutr.* **2018**, *37*, 2337–2343. [CrossRef] [PubMed]
13. De Cloet, J.; Simal, I.; Benoot, K.; Goossens, L. Assessment of the Adherence to ESPGHAN 2018 Guidelines in the Neonatal Intensive Care Unit of the Ghent University Hospital: A Retrospective Study. *Nutrients* **2023**, *15*, 2324. [CrossRef] [PubMed]
14. Embleton, N.D.; Moltu, S.J.; Lapillonne, A.; van den Akker, C.H.P.; Carnielli, V.; Fusch, C.; Gerasimidis, K.; van Goudoever, J.B.; Haiden, N.; Iacobelli, S.; et al. Enteral Nutrition in Preterm Infants (2022): A Position Paper From the ESPGHAN Committee on Nutrition and Invited Experts. *J. Pediatr. Gastroenterol. Nutr.* **2023**, *76*, 248–268. [CrossRef] [PubMed]
15. Lapillonne, A.; Kermorvant-Duchemin, E. A systematic review of practice surveys on parenteral nutrition for preterm infants. *J. Nutr.* **2013**, *143*, 2061S–2065S. [CrossRef]
16. Lapillonne, A.; Fellous, L.; Mokthari, M.; Kermorvant-Duchemin, E. Parenteral Nutrition Objectives for Very Low Birth Weight Infants: Results of a National Survey. *J. Pediatr. Gastroenterol. Nutr.* **2009**, *48*, 618–626. [CrossRef]
17. Taiwan Society of Neonatology. *Recommendation on Nutritional Care of Taiwan Preterm Infants*, 4th ed.; Taiwan Society of Neonatology: Taichung City, Taiwan, 2022.
18. Committee on Fetus and Newborn; Barfield, W.D.; Papile, L.A.; Baley, J.E.; Benitz, W.; Cummings, J.; Carlo, W.A.; Kumar, P.; Polin, R.A.; Tan, R.C.; et al. Levels of neonatal care. *Pediatrics* **2012**, *130*, 587–597.
19. Darmstadt, G.L.; Al Jaifi, N.H.; Arif, S.; Bahl, R.; Blennow, M.; Cavallera, V.; Chou, D.; Chou, R.; Comrie-Thomson, L.; Edmond, K. New World Health Organization recommendations for care of preterm or low birth weight infants: Health policy. *EClinicalMedicine* **2023**, *63*, 102155. [CrossRef]
20. Health Promotion Administration; Ministry of Health and Welfare, R.O.C. 2021 Statistics of Birth Reporting System. 2022. Available online: https://www.hpa.gov.tw/File/Attach/16584/File_20499.pdf (accessed on 17 September 2024).
21. Van Den Akker, C.H.; Te Braake, F.W.; Wattimena, D.J.; Voortman, G.; Schierbeek, H.; Vermes, A.; Van Goudoever, J.B. Effects of early amino acid administration on leucine and glucose kinetics in premature infants. *Pediatr. Res.* **2006**, *59*, 732–735. [CrossRef]
22. Blanco, C.L.; Gong, A.K.; Schoolfield, J.; Green, B.K.; Daniels, W.; Liechty, E.A.; Ramamurthy, R. Impact of early and high amino acid supplementation on ELBW infants at 2 years. *J. Pediatr. Gastroenterol. Nutr.* **2012**, *54*, 601–607. [CrossRef]

23. Valentine, C.; Fernandez, S.; Rogers, L.; Gulati, P.; Hayes, J.; Lore, P.; Puthoff, T.; Dumm, M.; Jones, A.; Collins, K. Early amino-acid administration improves preterm infant weight. *J. Perinatol.* **2009**, *29*, 428–432. [CrossRef] [PubMed]
24. Koletzko, B.; Goulet, O.; Hunt, J.; Krohn, K.; Shamir, R.; Group, P.N.G.W. 1. Guidelines on paediatric parenteral nutrition of the European Society of Paediatric Gastroenterology, Hepatology and Nutrition (ESPGHAN) and the European Society for Clinical Nutrition and Metabolism (ESPEN), supported by the European Society of Paediatric Research (ESPR). *J. Pediatr. Gastroenterol. Nutr.* **2005**, *41*, S1–S4. [PubMed]
25. Burattini, I.; Bellagamba, M.P.; Spagnoli, C.; D'Ascenzo, R.; Mazzoni, N.; Peretti, A.; Cogo, P.E.; Carnielli, V.P.; Network, M.N. Targeting 2.5 versus 4 g/kg/day of amino acids for extremely low birth weight infants: A randomized clinical trial. *J. Pediatr.* **2013**, *163*, 1278–1282.e1. [CrossRef] [PubMed]
26. Bloomfield, F.H.; Jiang, Y.; Harding, J.E.; Crowther, C.A.; Cormack, B.E.; Pro, V.T.G. Early Amino Acids in Extremely Preterm Infants and Neurodisability at 2 Years. *N. Engl. J. Med.* **2022**, *387*, 1661–1672. [CrossRef] [PubMed]
27. Alsweiler, J.M.; Kuschel, C.A.; Bloomfield, F.H. Survey of the management of neonatal hyperglycaemia in Australasia. *J. Paediatr. Child Health* **2007**, *43*, 632–635. [CrossRef]
28. Ramel, S.; Long, J.; Gray, H.; Durrwachter-Erno, K.; Demerath, E.; Rao, R. Neonatal hyperglycemia and diminished long-term growth in very low birth weight preterm infants. *J. Perinatol.* **2013**, *33*, 882–886. [CrossRef]
29. Auerbach, A.; Eventov-Friedman, S.; Arad, I.; Peleg, O.; Bdolah-Abram, T.; Bar-Oz, B.; Zangen, D.H. Long duration of hyperglycemia in the first 96 hours of life is associated with severe intraventricular hemorrhage in preterm infants. *J. Pediatr.* **2013**, *163*, 388–393. [CrossRef]
30. van der Lugt, N.M.; Smits-Wintjens, V.E.; van Zwieten, P.H.; Walther, F.J. Short and long term outcome of neonatal hyperglycemia in very preterm infants: A retrospective follow-up study. *BMC Pediatr.* **2010**, *10*, 52. [CrossRef]
31. Beardsall, K.; Vanhaesebrouck, S.; Ogilvy-Stuart, A.L.; Vanhole, C.; Palmer, C.R.; Ong, K.; Midgley, P.; Thompson, M.; Thio, M.; Cornette, L. Prevalence and determinants of hyperglycemia in very low birth weight infants: Cohort analyses of the NIRTURE study. *J. Pediatr.* **2010**, *157*, 715–719.e3. [CrossRef]
32. Agus, M.S.; Steil, G.M.; Wypij, D.; Costello, J.M.; Laussen, P.C.; Langer, M.; Alexander, J.L.; Scoppettuolo, L.A.; Pigula, F.A.; Charpie, J.R. Tight glycemic control versus standard care after pediatric cardiac surgery. *N. Engl. J. Med.* **2012**, *367*, 1208–1219. [CrossRef]
33. Macrae, D.; Grieve, R.; Allen, E.; Sadique, Z.; Morris, K.; Pappachan, J.; Parslow, R.; Tasker, R.C.; Elbourne, D. A randomized trial of hyperglycemic control in pediatric intensive care. *N. Engl. J. Med.* **2014**, *370*, 107–118. [CrossRef]
34. Beardsall, K.; Vanhaesebrouck, S.; Ogilvy-Stuart, A.L.; Vanhole, C.; Palmer, C.R.; van Weissenbruch, M.; Midgley, P.; Thompson, M.; Thio, M.; Cornette, L. Early insulin therapy in very-low-birth-weight infants. *N. Engl. J. Med.* **2008**, *359*, 1873–1884. [CrossRef] [PubMed]
35. Gleason, C.A.; Juul, S.E. *Avery's Diseases of the Newborn e-Book*; Elsevier Health Sciences: Amsterdam, The Netherlands, 2017.
36. Larchet, M.; Garabédian, M.; Bourdeau, A.; Gorski, A.-M.; Goulet, O.; Ricour, C. Calcium metabolism in children during long-term total parenteral nutrition: The influence of calcium, phosphorus, and vitamin D intakes. *J. Pediatr. Gastroenterol. Nutr.* **1991**, *13*, 367–375. [PubMed]
37. Mutanen, A.; Mäkitie, O.; Pakarinen, M.P. Risk of metabolic bone disease is increased both during and after weaning off parenteral nutrition in pediatric intestinal failure. *Horm. Res. Paediatr.* **2013**, *79*, 227–235. [CrossRef] [PubMed]
38. Appleman, S.S.; Kalkwarf, H.J.; Dwivedi, A.; Heubi, J.E. Bone deficits in parenteral nutrition-dependent infants and children with intestinal failure are attenuated when accounting for slower growth. *J. Pediatr. Gastroenterol. Nutr.* **2013**, *57*, 124. [CrossRef] [PubMed]
39. Watrobska-Swietlikowska, D. Compatibility of maximum inorganic and organic calcium and phosphate content in neonatal parenteral solutions. *Sci. Rep.* **2019**, *9*, 10525. [CrossRef]
40. Wang, H.-J.; Hsieh, Y.-T.; Liu, L.-Y.; Huang, C.-F.; Lin, S.-C.; Tsao, P.-N.; Chou, H.-C.; Yen, T.-A.; Chen, C.-Y. Use of sodium glycerophosphate in neonatal parenteral nutrition solutions to increase calcium and phosphate compatibility for preterm infants. *Pediatr. Neonatol.* **2020**, *61*, 331–337. [CrossRef]
41. Hsu, P.C.; Tsao, P.N.; Chou, H.C.; Huang, H.C.; Yen, T.A.; Chen, C.Y. Sodium glycerophosphate use in parenteral nutrition improves mineral metabolism in extremely low birth weight infants. *J. Pediatr.* **2023**, *253*, 63–71. [CrossRef]
42. Leaf, A. Early enteral feeding in high-risk preterm infants. *Infant* **2007**, *3*, 27–30.
43. Maroulis, J.; Kalfarentzos, F. Complications of parenteral nutrition at the end of the century. *Clin. Nutr.* **2000**, *19*, 295–304. [CrossRef]
44. Dorling, J.; Abbott, J.; Berrington, J.; Bosiak, B.; Bowler, U.; Boyle, E.; Embleton, N.; Hewer, O.; Johnson, S.; Juszczak, E. Controlled trial of two incremental milk-feeding rates in preterm infants. *N. Engl. J. Med.* **2019**, *381*, 1434–1443. [CrossRef] [PubMed]
45. Yang, W.-C.; Fogel, A.; Lauria, M.E.; Ferguson, K.; Smith, E.R. Fast Feed Advancement for Preterm and Low Birth Weight Infants: A Systematic Review and Meta-analysis. *Pediatrics* **2022**, *150*, e2022057092G. [CrossRef]
46. Alsaied, A.; Islam, N.; Thalib, L. Global incidence of necrotizing enterocolitis: A systematic review and meta-analysis. *BMC Pediatr.* **2020**, *20*, 344. [CrossRef] [PubMed]
47. Wittwer, A.; Hascoet, J.M. Impact of introducing a standardized nutrition protocol on very premature infants' growth and morbidity. *PLoS ONE* **2020**, *15*, e0232659. [CrossRef] [PubMed]

48. Faienza, M.F.; D'Amato, E.; Natale, M.P.; Grano, M.; Chiarito, M.; Brunetti, G.; D'Amato, G. Metabolic Bone Disease of Prematurity: Diagnosis and Management. *Front. Pediatr.* **2019**, *7*, 143. [CrossRef] [PubMed]
49. Chan, G.; Armstrong, C.; Moyer-Mileur, L.; Hoff, C. Growth and bone mineralization in children born prematurely. *J. Perinatol.* **2008**, *28*, 619–623. [CrossRef]
50. Mitchell, S.M.; Rogers, S.P.; Hicks, P.D.; Hawthorne, K.M.; Parker, B.R.; Abrams, S.A. High frequencies of elevated alkaline phosphatase activity and rickets exist in extremely low birth weight infants despite current nutritional support. *BMC Pediatr.* **2009**, *9*, 47. [CrossRef]
51. Catache, M.; Leone, C. Role of plasma and urinary calcium and phosphorus measurements in early detection of phosphorus deficiency in very low birthweight infants. *Acta Paediatr.* **2003**, *92*, 76–80. [CrossRef]
52. Backström, M.; Kouri, T.; Kuusela, A.L.; Sievänen, H.; Koivisto, A.M.; Ikonen, R.; Mäki, M. Bone isoenzyme of serum alkaline phosphatase and serum inorganic phosphate in metabolic bone disease of prematurity. *Acta Paediatr.* **2000**, *89*, 867–873. [CrossRef]
53. Hung, Y.L.; Chen, P.C.; Jeng, S.F.; Hsieh, C.J.; Peng, S.S.F.; Yen, R.F.; Chou, H.C.; Chen, C.Y.; Tsao, P.N.; Hsieh, W.S. Serial measurements of serum alkaline phosphatase for early prediction of osteopaenia in preterm infants. *J. Paediatr. Child Health* **2011**, *47*, 134–139. [CrossRef]
54. Chinoy, A.; Mughal, M.Z.; Padidela, R. Metabolic bone disease of prematurity: Causes, recognition, prevention, treatment and long-term consequences. *Arch. Dis. Child.-Fetal Neonatal Ed.* **2019**, *104*, F560–F566. [CrossRef] [PubMed]
55. Kuzma-O'Reilly, B.; Duenas, M.L.; Greecher, C.; Kimberlin, L.; Mujsce, D.; Miller, D.; Walker, D.J. Evaluation, development, and implementation of potentially better practices in neonatal intensive care nutrition. *Pediatrics* **2003**, *111*, e461–e470. [CrossRef] [PubMed]

nutrients

MDPI

Article

Incidence and Risk Factors of Refeeding Syndrome in Preterm Infants

Suzan S. Asfour [1], Belal Alshaikh [2], Maya Mathew [1], Dina I. Fouda [1] and Mountasser M. Al-Mouqdad [3,*]

1 Clinical Pharmacy Department, Pharmaceutical Care Services, King Saud Medical City, Riyadh 12746, Saudi Arabia; asfsuzan@gmail.com (S.S.A.)
2 Department of Pediatrics, Cumming School of Medicine, University of Calgary, Calgary, AB T2N 1N4, Canada
3 Neonatal Intensive Care, Hospital of Pediatrics, King Saud Medical City, Riyadh 12746, Saudi Arabia
* Correspondence: m.almouqdad@ksmc.med.sa; Tel.: +966-114355555 (ext. 506); Fax: +966-114354406

Abstract: This study aimed to evaluate the incidence and risk factors associated with refeeding syndrome (RFS) in preterm infants (≤32 weeks gestational age) during their first week of life. Infants (gestational age ≤ 32 weeks; birth weight < 1500 g) who were admitted to the neonatal intensive care unit (NICU), level III, and received parenteral nutrition between January 2015 and April 2024 were retrospectively evaluated. Modified log-Poisson regression with generalized linear models and a robust variance estimator was applied to adjust the relative risk of risk factors. Of the 760 infants identified, 289 (38%) developed RFS. In the multivariable regression analysis, male, intraventricular hemorrhage (IVH), and sodium phosphate significantly affected RFS. Male infants had significantly increased RFS risk (aRR1.31; 95% CI 1.08–1.59). The RFS risk was significantly higher in infants with IVH (aRR 1.71; 95% CI 1.27–2.13). However, infants who received higher sodium phosphate in their first week of life had significantly lower RFS risk (aRR 0.67; 95% 0.47–0.98). This study revealed a notable incidence of RFS among preterm infants aged ≤32 gestational weeks, with sex, IVH, and low sodium phosphate as significant risk factors. Refined RFS diagnostic criteria and targeted interventions are needed for optimal management.

Keywords: parenteral nutrition; refeeding syndrome; sodium phosphate; preterm infants

1. Introduction

Despite the lack of a universally accepted definition of refeeding syndrome (RFS), most studies agree that it is characterized by fluid and electrolyte imbalances occurring when nutrition is reintroduced after prolonged malnutrition or starvation [1,2]. RFS can result from both enteral nutrition and parenteral nutrition (PN), and it often involves significant electrolyte disturbances, particularly phosphorus, and can lead to severe complications affecting multiple organ systems, potentially resulting in death [1–3].

Studies have found that adult patients with conditions including tuberculosis, cancer, psychiatric disorders, and chronic diseases who are malnourished are at risk of developing RFS [4,5]. Consequently, these patients require a longer hospital stay and have a higher mortality risk during hospitalization [6–8]. RFS was first recognized in adult patients who experienced severe malnutrition or starvation, such as prisoners of war, individuals with anorexia nervosa, or those recovering from famine [4,5,9]. Upon refeeding, these individuals developed potentially life-threatening symptoms. The underlying mechanism involves a metabolic shift from a catabolic state, where the body primarily uses fat and protein for energy, to an anabolic state, where carbohydrate intake stimulates insulin release. Insulin promotes the cellular uptake of glucose, potassium, magnesium, and phosphate. This sudden shift can lead to profound hypophosphatemia, as phosphate is rapidly utilized in cells for ATP production and other metabolic processes [10]. The resultant electrolyte imbalance, particularly low phosphate levels, can cause severe complications, including cardiac, respiratory, and neurological dysfunction [6–8]. RFS is also observed in the pediatric population, particularly among those admitted to a pediatric intensive care unit (PICU) [11]. Up to 25% of children admitted to a PICU are malnourished, placing them at a heightened risk of RFS development if nutrition is delayed for >5 days [11–13]. A study reported a mortality rate of up to 6% among children with RFS [14].

Neonates were believed to be not at risk of RFS because they received enteral nutrition or PN shortly after birth. However, neonates, particularly those who are extremely premature, small for gestational age (SGA), affected by intrauterine growth restriction (IUGR), or have very low birth weight (VLBW), can become malnourished at a very early stage of life, putting them at risk for RFS [2].

The incidence of RFS in the neonatal population varies widely because of inconsistencies in its definition. This variability stems from whether the diagnosis is based solely on hypophosphatemia or includes hypokalemia and hypomagnesemia. The varied definition of hypophosphatemia and differences in inclusion criteria across studies further contributed to the wide range in the reported incidence rates [15–20].

Several studies have identified fetal growth restriction, irrespective of its cause—placental insufficiency or genetic disorders—as a significant risk factor for RFS development in newborns [15,19]. Other studies have highlighted that being SGA is also associated with an increased RFS risk [16,19,20]. Furthermore, reports have suggested that the early introduction of high amino acid content through PN in premature infants carries a substantial risk of RFS [16,21].

Despite these findings, these studies share several limitations. They often have small sample sizes, include both preterm and term infants, and utilize varying definitions of RFS and hypophosphatemia [17–19,22]. Moreover, many of these studies do not report the concentrations of other PN components, such as dextrose and proteins [15,17,23].

These factors make it challenging to ascertain the true incidence rate and risk factors of RFS in preterm babies. Therefore, this study aimed to determine the incidence and risk factors for RFS development in premature infants aged ≤32 gestational weeks during the first week of life.

2. Materials and Methods

2.1. Study Design

This retrospective chart review included preterm infants who were admitted to the neonatal intensive care unit (NICU) of the King Saud Medical City (KSMC), a tertiary referral center, between January 2015 and June 2024. The NICU, level III, at KSMC has an average annual admission of 1100 patients.

This study was conducted in accordance with the Declaration of Helsinki and Good Pharmacoepidemiology Practice Guidelines and was approved by the Medical Ethical Review Committee of KSMC (Reference number H1RI-12-May24-01), which also waived the need for informed consent.

2.2. Inclusion and Exclusion Criteria

Very preterm infants who weighed <1500 g at birth, were born at KSMC at ≤32 weeks of gestation, and were admitted to the NICU, level III, were included. All the included newborns received PN plus lipid emulsion within the first 24 h of life. Infants with a known chromosomal or genetic abnormality and significant congenital defects or congenital infections, did not receive PN, were not born at KSMC or were transferred to another hospital or died within the first 7 days, and had nonretrievable data were excluded from the analysis.

2.3. Data Collection and Follow-Up

Neonatal data from NICU admission until discharge or death were retrieved. Demographic data and clinical and outcome data, including major morbidities associated with prematurity, were also reviewed and collected. Maternal data, including antenatal steroid treatment, mode of delivery, and presence of gestational diabetes mellitus and maternal hypertension, were also obtained.

2.4. Study Outcome

The primary outcomes of this study were the incidence of RFS in the first 7 days after birth and risk factors of RFS in very preterm infants.

2.5. Definitions

Nutrition Protocol

PN: PN was started early after birth using starter PN. Individualized PN was prescribed daily. Starter PN contains 10% dextrose, 4% amino acids, and 0.01 mmol/mL calcium gluconate [2]. Individualized PN solution containing amino acids (3.5–4 g/kg/day), dextrose (5–12 mg/kg/min), lipid emulsion (1–3 g/kg/day), sodium chloride (1–3 mmol/kg/day), sodium acetate (1–2 mmol/kg/day), sodium phosphate (1–2 mmol/kg/day), potassium chloride (1–3 mmol/kg/day), potassium acetate (1–2 mmol/kg/day), potassium phosphate (1–2 mmol/kg/day), trace elements (Peditrace®), and water- and fat-soluble vitamins (Soluvit® N and Vitalipid® N Infant, respectively) was started within the first 24 h of life and infused continuously for 24 h [2].

RFS: A clear definition of neonatal RFS has not yet been established: hypercalcemia (>2.8 mmol.L^{-1}), hypophosphatemia (between >1.1 and <1.6 mmol.L^{-1}), and severe hypophosphatemia (<1.0 mmol.L^{-1}) [16,22,24–26].

2.6. Statistical Analysis

Before the analysis, the dataset was reviewed and checked for missing data. Data were analyzed using IBM SPSS Statistics for Windows version 25.0 (IBM Corp., Armonk, NY, USA).

Infant and maternal variables were presented using descriptive statistics, including median, interquartile range (IQR), frequency, and percentage. The Mann–Whitney U test was used for between-group comparisons of ordinal qualitative variables. Fisher's exact test was utilized to determine the association between categorical variables. For between-group comparisons of continuous variables, the unpaired Student's *t*-test was used for normally distributed data, whereas the Mann–Whitney U test was used for non-normally distributed data. The Kolmogorov–Smirnov test and a visual inspection of histograms were performed to evaluate the distribution of quantitative variables.

To determine the risk factors of RFS in premature infants, a univariate relative risk analysis on the recorded variables (gestational age, birth weight, SGA, delivery mode, sex, 1 and 5 min Apgar scores, maternal hypertension, antenatal steroid treatment, premature rupture of the membrane, gestational diabetes mellitus, necrotizing enterocolitis, surfactant use, late-onset sepsis, dextrose intake, amino acids, lipid emulsion, sodium chloride, sodium acetate, sodium phosphate, potassium chloride, potassium acetate, potassium phosphate, magnesium sulfate, calcium gluconate, and calcium-to-phosphate ratio) was first performed because they were considered potential confounders. All factors with a *p*-value of <0.05 in the univariate analysis were included in the final multivariable regression model. Modified log-Poisson regression with generalized linear models and a robust variance estimator (Huber–White) were applied for univariate relative risk analysis and to the models to adjust the relative risk for RFS risk factors in premature infants. All statistical tests were two-tailed, and *p*-values of <0.05 were considered significant.

3. Results

During the study period, 2465 preterm infants with ≤32 gestational weeks and birth weight < 1500 g were admitted to the level 3 NICU. Of them, 760 met the inclusion criteria and were eligible for inclusion in the final analysis (Figure 1).

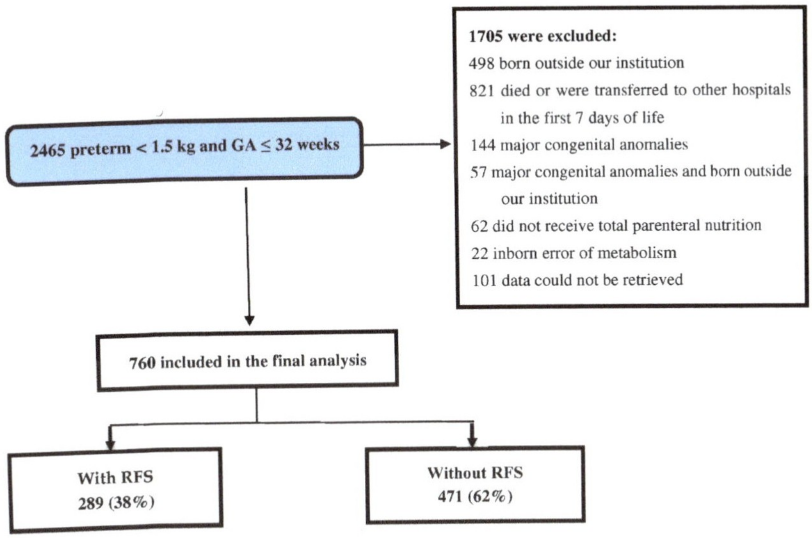

Figure 1. Flow chart of patient selection. GA: gestational age.

Among the 760 infants, RFS developed in 289 (38%). Maternal and neonatal characteristics are summarized in Table 1. Figure 2 shows the percentages of RFS, hypercalcemia, hypophosphatemia, and severe hypophosphatemia. Infants with RFS have lower gestational age, birthweight, and length and head circumference than infants without RFS ($p < 0.001$). The Apgar scores at 1 and 5 min were also lower in infants with RFS than in those without RFS ($p < 0.001$). In addition, male infants were at higher risk of RFS than female infants ($p < 0.001$). Infants with RFS received more surfactant and required mechanical ventilators more often than those without RFS ($p < 0.001$). Moreover, infants with RFS had more severe IVH than infants without RFS ($p < 0.001$).

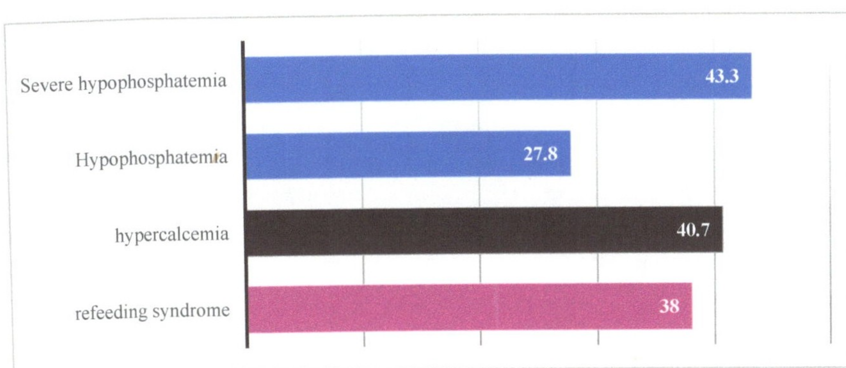

Figure 2. Percentages of refeeding syndrome, hypercalcemia, hypophosphatemia, and severe hypophosphatemia ($n = 760$).

In the univariate analysis, the median intake of macronutrients, including parenteral lipids (g/kg/day) and amino acids (g/kg/day), in the first postnatal week was significantly higher in infants with RFS than in those without RFS ($p < 0.001$). TPN duration was also significantly higher in infants with RFS than in those without it ($p = 0.002$) (Table 2).

Table 1. Demographic characteristics of the mothers and infants with or without refeeding syndrome (RFS, *n* = 760).

Variable	ALL				Gestational Age < 28 Weeks			
	N	without RFS (*n* = 471)	with RFS (*n* = 289)	*p*-Value	N	without RFS (*n* = 128)	with RFS (*n* = 136)	*p*-Value
Gestational age (weeks), median (IQR)	760	29 (27.0–31.0)	28 (26–30.0)	**<0.001** *	264	26 (25.0–27.0)	26 (25.0–27.0)	**0.02** *
Birth weight (grams) (IQR)	760	1180 (950–1370)	950 (765–1200)	**<0.001** *	264	845 (711.25–960)	775 (661.25–905)	**0.008** *
Length (cm), median (IQR)	760	38 (35–40)	35 (32–38)	**<0.001** *	264	33 (31–35)	33 (31–35)	0.10
Head circumference (cm) (IQR)	760	27 (25–28)	25 (23–27)	**<0.001** *	264	24 (23–25)	23 (22–25)	**0.03** *
Small for gestational age, *n* (%)	760	63 (13.4)	47 (16.3)	0.29	264	8 (6.3)	9 (6.6)	1
1 min Apgar score, median (IQR)	760	6 (4–7)	5 (3–6)	**<0.001** *	264	5 (3–6)	4 (2–6)	**0.03** *
5 min Apgar score, median (IQR)	760	7 (6–8)	7 (7–8)	**<0.001** *	264	5 (6–7)	4 (6–7)	0.17
Male, *n* (%)	760	221 (46.9)	174 (60.2)	**<0.001** *	264	70 (54.7)	90 (66.2)	0.06
Antenatal steroid treatment, *n* (%)	760	253 (53.7)	147 (50.9)	0.45	264	70 (54.7)	71 (52.2)	0.71
Gestational diabetes mellitus, *n* (%)	760	27 (5.7)	14 (4.8)	0.37	264	7 (5.5)	4 (2.9)	0.36
Maternal hypertension, *n* (%)	760	119 (25.3)	67 (23.2)	0.54	264	26 (20.3)	27 (19.9)	1.0
Preterm rupture of membrane, *n* (%)	760	59 (12.5)	20 (6.9)	**0.01** *	264	17 (13.3)	12 (8.8)	0.33
Cesarean section, *n* (%)	760	223 (47.3)	146 (50.5)	0.41	264	72 (56.3)	75 (55.1)	0.90
Expressed breast milk, *n* (%)	760	247 (52.4)	145 (50.2)	0.55	264	75 (58.6)	64 (47.1)	0.06
Respiratory distress syndrome required surfactant, *n* (%)	760	281 (59.7)	224 (77.5)	**<0.001** *	264	120 (93.8)	128 (94.1)	1
Mechanical ventilation, *n* (%)	760	305 (64.8)	242 (83.7)	**<0.001** *	264	122 (95.3)	132 (97.1)	0.53
Patent ductus arteriosus requiring treatment, *n* (%)	760	35 (7.4)	34 (11.8)	0.05	264	24 (18.8)	28 (20.6)	0.76
Intraventricular hemorrhage, *n* (%)	760	124 (26.3)	157 (54.5)	**<0.001** *	264	64 (50)	96 (70.6)	**0.001** *
Intraventricular hemorrhage grades 3 and 4, *n* (%)	760	63 (13.3)	74 (25.6)	**<0.001** *	264	41 (32)	56 (41.2)	0.13

* *p*-values < 0.05.

Table 2. Nutritional and electrolyte supplementation characteristics of infants with or without refeeding syndrome (RFS, *n* = 760).

Variables	ALL				Gestational Age < 28 Weeks			
	N	without RFS (*n* = 471)	with RFS (*n* = 289)	*p*-Value	N	without RFS (*n* = 128)	with RFS (*n* = 136)	*p*-Value
Average parenteral lipid intake in the 1st 7 days (g/kg/day), median (IQR)	760	2 (1.54–2.35)	2.1 (1.67–2.5)	**0.02** *	264	1.8 (1.4–2.3)	1.8 (1.3–2.4)	0.95
Average parenteral amino acid intake in the 1st 7 days (g/kg/day), median (IQR)	760	3.90 (3.67–4.0)	3.97 (3.75–4.0)	**0.01** *	264	4 (3.7–4.0)	4 (3.8–4.0)	0.13
Average parenteral dextrose intake in the 1st 7 days (mg/kg/min), median (IQR)	760	8.47 (7.74–9.12)	8.46 (7.65–9.04)	0.55	264	8 (7–8.8)	7.7 (6.4–8.5)	**0.03** *
Average parenteral phosphate intake in the 1st 7 days (mg/kg/min), median (IQR)	760	0.30 (0–0.6)	0.24 (0–0.4)	**0.01** *	264	0.30 (0.02–0.62)	0.2 (0–0.44)	**0.03** *
PN duration, median (IQR)	760	14 (7–29)	18 (9–35)	**0.002** *	264	28 (12–48)	23 (11–41)	0.45

Table 2. *Cont.*

Variables	ALL				Gestational Age < 28 Weeks			
	N	without RFS (n = 471)	with RFS (n = 289)	p-Value	N	without RFS (n = 128)	with RFS (n = 136)	p-Value
Average magnesium sulfate intake in the 1st 7 days (mmol/kg/day), median (IQR)	760	0.11 (0.07–0.13)	0.11 (0.07–0.13)	0.60	264	0.10 (0.07–0.13)	0.11 (0.07–0.14)	0.24
Average calcium gluconate intake in the 1st 7 days (mmol/kg/day), median (IQR)	760	0.71 (0.57–0.92)	0.79 (0.61–1)	**0.001** *	264	0.77 (0.64–0.96)	0.86 (0.64–1)	0.15
Average sodium chloride intake in the 1st 7 days (mmol/kg/day), median (IQR)	760	0.14 (0–0.60)	0 (0–0.40)	**<0.001** *	264	0.0 (0.0–0.43)	0.0 (0.0–0.29)	0.08
Average sodium acetate intake in the 1st 7 days (mmol/kg/day), median (IQR)	760	0.57 (0–1.14)	0.43 (0–0.96)	0.08	264	0.86 (0.104–1.14)	0.57 (0.0–1.0)	**0.02** *
Average sodium Phosphate intake in the 1st 7 days (mmol/kg/day), median (IQR)	760	0.37 (0.14–0.60)	0.29 (0–0.50)	**<0.001** *	264	0.3 (0.1–0.59)	0.2 (0.0–0.43)	**0.001** *
Average potassium chloride intake in the 1st 7 days (mmol/kg/day), median (IQR)	760	0.4 (0.0–1.0)	0.2 (0.0–0.8)	**0.001** *	264	0.4 (0.0–1.0)	0.0 (0.0–0.6)	**0.001** *
Average potassium acetate intake in the 1st 7 days (mmol/kg/day), median (IQR)	760	0 (0–0.43)	0.14 (0–0.71)	**<0.001** *	264	0.14 (0–0.64)	0.29 (0–0.86)	0.11
Average potassium phosphate intake in the 1st 7 days (mmol/kg/day), median (IQR)	760	0 (0–0.38)	0.1 (0.0–0.38)	0.14	264	0.0 (0.0–0.42)	0.1 (0.0–0.40)	0.14
Average calcium-to-phosphate ratio intake in the 1st 7 days (mmol/kg/day), median (IQR)	760	0.61 (0.38–0.90)	0.87 (0.54–1.44)	**<0.001** *	264	0.65 (0.42–0.99)	0.97 (0.69–1.68)	**<0.001** *

* p-values < 0.05.

Interestingly, the median intake of sodium chloride, sodium phosphate, and potassium chloride in the first postnatal week was significantly higher in infants without RFS than in those with RFS (p < 0.001, <0.001, 0.001, respectively) (Table 2).

In contrast, the median intake of calcium gluconate and potassium acetate and the calcium-to-phosphate ratio in the first postnatal week were significantly higher in infants with RFS than in those without RFS (p = 0.001, <0.001, <0.001, respectively) (Table 2).

The multivariable regression analysis performed after adjusting the variables that were significant in the univariate analysis revealed that male sex, IVH, and sodium phosphate significantly affect RFS risk. Male infants had significantly increased RFS risk (aRR1.31; 95% CI1.08–1.59). The RFS risk was significantly higher in infants with and without IVH (aRR 1.71; 95% CI 1.27–2.13). However, infants who received higher sodium phosphate in their first week of life had a significantly lower RFS risk (aRR 0.67; 95% 0.47–0.98) (Table 3).

Table 3. Univariate and multivariable regression analyses of risk factors of refeeding syndrome (RFS) in very preterm infants.

Risk Factor	All				Gestational Age < 28 Weeks			
	Unadjusted RR 95% CI	p-Value	Adjusted RR 95% CI	p-Value	Unadjusted RR 95% CI	p-Value	Adjusted RR 95% CI	p-Value
Gestational age	0.89 (0.86–0.92)	**<0.001** *	1.02 (0.95–1.09)	0.59	0.90 (0.83–0.99)	**0.03** *	1.0 (0.89–1.13)	0.98
Birth weight	0.99 (0.98–1.0)	**<0.001** *	0.99 (0.98–1.0)	0.07	0.99 (0.98–1)	**0.01** *	1.0 (0.99–1.001)	0.63

Table 3. *Cont.*

Risk Factor	All				Gestational Age < 28 Weeks			
	Unadjusted RR 95% CI	***p*-Value**	**Adjusted RR 95% CI**	***p*-Value**	**Unadjusted RR 95% CI**	***p*-Value**	**Adjusted RR 95% CI**	***p*-Value**
Small for gestational age	1.15 (0.90–1.46)	0.25	–	–	0.97 (0.61–1.55)	0.90	–	–
Male	1.40 (1.16–1.69)	**<0.001** *	1.31 (1.08–1.59)	**0.006** *	0.78 (0.61–1.01)	0.06	–	–
Cesarean section	1.08 (0.90–1.30)	0.40	–	–	1.02 (0.81–1.29)	0.86	–	–
1 min Apgar score	0.89 (0.85–0.93)	**<0.001** *	0.98 (0.91–1.04)	0.48	0.93 (0.88–0.99)	**0.03** *	0.99 (0.93–1.06)	0.72
5 min Apgar score	0.88 (0.82–0.94)	**<0.001** *	1.04 (0.94–1.14)	0.47	0.95 (0.86–1.04)	0.28	–	–
Surfactant	1.74 (1.38–2.19)	**<0.001** *	0.81 (0.58–1.12)	0.20	0.97 (058–1.61)	0.90	–	–
Patent ductus requiring treatment	1.34 (1.03–1.73)	**0.03** *	0.84 (0.64–1.10)	0.21	1.06 (0.79–1.40)	0.70	–	–
Intraventricular hemorrhage	2.03 (1.70–2.44)	**<0.001** *	1.71 (1.27–2.13)	**<0.001** *	1.56 (1.19–2.05)	**0.001** *	0.75 (0.57–1.0)	0.05
Nutritional and anthropometrics								
Average amino acid intake	1.38 (1.02–1.88)	**0.03** *	1.29 (0.93–1.81)	0.13	1.60 (0.98–2.59)	0.06	–	–
Average dextrose intake	0.95 (0.89–1.02)	0.16	–	–	0.91 (0.85–0.99)	**0.02**	0.96 (0.89–1.04)	0.32
Average lipid intake	0.88 (0.76–1.03)	0.11	–	–	0.99 (0.82–1.20)	0.9	–	–
Average magnesium sulfate intake	1.19 (0.16–8.81)	0.86	–	–	2.36 (0.16–33.78)	0.53	–	–
Average calcium gluconate intake	1.81 (1.28–2.54)	**0.001** *	1.29 (0.91–1.82)	0.15	1.30 (0.83–2.02)	0.25	–	–
Average sodium chloride intake	0.64 (0.50–0.82)	**<0.001** *	0.86 (0.68–1.09)	0.21	0.81 (0.57–1.14)	0.23	–	–
Average sodium acetate intake	0.86 (0.73–1.01)	0.07	–	–	0.82 (0.66–1.02)	0.08	–	–
Average sodium phosphate intake	0.55 (0.37–0.72)	**<0.001** *	0.67 (0.47–0.98)	**0.03** *	0.50 (0.32–0.78)	**0.002** *	0.60 (0.38–0.96)	**0.03** *
Average potassium chloride intake	0.82 (0.69–0.97)	**0.02** *	0.98 (0.83–1.17)	0.84	0.75 (0.55–1.01)	0.06	–	–
Average potassium acetate intake	1.40 (1.17–1.67)	**<0.001** *	1.05 (0.86–1.29)	0.61	1.17 (0.94–1.46)	0.17	–	–
Average potassium phosphate intake	1.22 (0.91–1.64)	0.18	–	–	1.23 (0.91–1.67)	0.18	–	–
Average calcium-to-phosphate ratio intake	1.13 (1.07–1.19)	**<0.001**	1.03 (0.99–1.07)	0.07	1.06 (1.02–1.10)	0.001	1.02 (0.98–1.06)	0.32

* *p*-values of <0.05.

Furthermore, the risk factors of RFS in neonates with gestational age < 28 weeks were analyzed. Among the 264 infants, RFS developed in 136 (51.5%). Infants with RFS had lower gestational age, birthweight, head circumference, and Apgar score at 1 min than infants without RFS ($p = 0.02$, 0.008, 0.03, 0.03, respectively) (Table 1).

Infants aged <28 gestational weeks with RFS received lower amounts of dextrose, sodium acetate, sodium phosphate, and potassium chloride than infants without RFS ($p = 0.03$, 0.02, 0.001, 0.001, respectively) (Table 2). In addition, the average calcium-to-phosphate ratio intake in the first week of life was higher in infants with RFS than in those without RFS ($p < 0.001$) (Table 2).

The multivariable regression analysis after adjusting the confounders revealed that infants who received higher amounts of sodium phosphate in the first week of life had a significantly lower RFS risk (aRR 0.60; 95% CI 0.38–0.96) (Table 3).

4. Discussion

This study examined the incidence rate and identified the risk factors associated with RFS in preterm infants. The findings revealed that the incidence of RFS was approximately 38%, highlighting the significant prevalence of this condition in this study population.

The incidence rates of RFS in the neonatal population vary widely, ranging from 20% to 90%. This variability can be attributed to several factors, including differences in the definition of RFS across studies. Some studies define RFS solely based on hypophosphatemia, whereas others include additional electrolyte imbalances such as hypokalemia or hypercalcemia [16–18,20,21,23,27]. In addition, the incidence rates are influenced by the prevalence of IUGR and SGA within the studied populations and the amounts of amino acids provided in PN during the first few days of life.

In this study, RFS was defined based on the ProVIDe Trial, which defines serum hypophosphatemia as <1.4 mmol/L and hypercalcemia as adjusted calcium >2.8 mmol/L [20,28]. The use of this standardized definition allows for more accurate comparisons of our findings with those of other studies and emphasizes the need for consistent criteria in evaluating the incidence and risk factors of RFS.

In this study, several key risk factors were found to be associated with RFS development in preterm infants during the first week of life. The findings indicate that sex, IVH, and lower sodium phosphate intake in the first week of life are significant risk factors for RFS.

Male infants were found to have a higher risk of RFS development than female infants. This finding is particularly intriguing because sex has not been widely recognized as a risk factor for RFS in the existing literature. The ProVIDe Trial is one of the few studies suggesting a potential link, noting a trend toward increased RFS incidence in male infants without definitively establishing sex as a risk factor [20]. Our findings support this observation, indicating that being of male sex may contribute to RFS susceptibility in preterm infants. The increased risk in male infants may be due to hormonal or genetic factors, although this hypothesis requires further investigation.

This study found a significant association between IVH and RFS development in preterm infants during their first week of life. This finding aligns with two other studies that reported a significant association between IVH and hypophosphatemia, a key component of RFS [20,23]. However, four other studies did not find a significant association [15,21,29,30]. This discrepancy in the findings could be attributed to differences in the overall incidence of IVH in study populations, sample sizes, and clinical protocols. The highest incidence of IVH in very premature infants typically occurs within the first week of life, coinciding with the period when RFS is most likely to develop. Given that IVH is a severe morbidity associated with prematurity, the significant association with RFS highlights the need for close monitoring and management of these infants during this critical period.

This study identified low sodium phosphate intake as a significant risk factor for RFS development in preterm infants, particularly those aged <28 weeks of gestation. Notably, our institution does not initiate sodium phosphate in PN within the first 2 days of life. Consequently, a lower incidence of RFS was noted in preterm infants who received higher amounts of phosphate over the first 7 days of life. These findings are consistent with those of Bustos-Lozano et al. and Mulla et al. [27,31]. Bustos-Lozano et al. reported hypophosphatemia in 29% of infants who received phosphate within their first 48 h of life compared with 69% in those who did not receive early phosphate supplementation. Similarly, Mulla et al. found hypophosphatemia in 60% of infants with a calcium-to-phosphate ratio of 1.3–1.5:1 compared with 35% in those with a 1:1 ratio. These results highlight the importance of early phosphate supplementation and adjusting the calcium-to-phosphate ratio to reduce the risk of RFS and its complications.

High amino acid intake was not identified as a risk factor for RFS development in our cohort. Our neonatal unit adheres to the recommendation of providing adequate protein immediately after birth, administering 3.5–4 g/kg/day to support growth and neurodevelopmental outcomes [32,33]. As a result, infants with and without RFS received the same amount of enteral and parenteral amino acids. We do not practice adjusting the rate of amino acid intake based on serum urea levels because this measure is not considered reliable in preterm infants.

This study has several limitations. First, the retrospective design and single tertiary center setting may limit the generalizability of the findings. Second, the association between IUGR and RFS could not be investigated because of a lack of antenatal care information for the majority of our infants' mothers, as many did not receive antenatal care or we did not have access to their antenatal care history.

Despite these limitations, this study has several strengths. The definition of RFS used in the ProVIDe Trial, a large multicenter prospective study with a substantial sample size, was adopted, which enhances the reliability and comparability of our results. Furthermore, this study has a robust sample size compared with many previous studies, allowing for more reliable statistical analysis. Moreover, we meticulously tracked changes in the components of PN between infants with and without RFS. These components included proteins, carbohydrates, and lipids, as well as micronutrients, such as magnesium and sodium and potassium supplements.

5. Conclusions

In our NICU population, the incidence of RFS is high, revealing significant clinical concerns. Some risk factors, such as sex, are beyond our control, whereas others, including nutritional protocols and strategies to minimize the risk of IVH, can be managed more effectively. This study highlights the necessity for further advanced prospective studies to accurately define neonatal RFS and determine optimal amino acid intake and the early initiation of phosphate supplementation. Although our findings contribute to the growing body of evidence, more studies are essential to develop more effective prevention and management strategies for preterm infants at risk of RFS, ultimately improving their outcomes and care.

Author Contributions: Conceptualization, S.S.A. and M.M.A.-M.; data curation, M.M. and D.I.F.; formal analysis, S.S.A.; methodology, B.A. and M.M.A.-M.; writing—original draft, S.S.A. and M.M.A.-M.; created the figures and tables, S.S.A.; writing—review and editing, S.S.A., B.A., M.M., D.I.F. and M.M.A.-M. All authors have read and agreed to the published version of the manuscript.

Funding: This research received no external funding.

Institutional Review Board Statement: This study was conducted in accordance with the Declaration of Helsinki and Good Pharmacoepidemiology Practice Guidelines and was approved by the Medical Ethical Review Committee of KSMC (reference number H1RI-12-May24-01, 12 May 2024).

Informed Consent Statement: The requirement for consent was waived.

Data Availability Statement: The data presented in this study are available on request from the corresponding author. The data is not available publicly due to Ethical reason.

Acknowledgments: We are particularly grateful to all the people who have given us help on our article.

Conflicts of Interest: The authors declare no conflicts of interest.

References

1. Skipper, A. Refeeding syndrome or refeeding hypophosphatemia: A systematic review of cases. *Nutr. Clin. Pract.* **2012**, *27*, 34–40. [CrossRef] [PubMed]

2. da Silva, J.S.V.; Seres, D.S.; Sabino, K.; Adams, S.C.; Berdahl, G.J.; Citty, S.W.; Cober, M.P.; Evans, D.C.; Greaves, J.R.; Gura, K.M.; et al. ASPEN consensus recommendations for refeeding syndrome. *Nutr. Clin. Pract.* **2020**, *35*, 178–195. [CrossRef] [PubMed]
3. McKnight, C.L.; Newberry, C.; Sarav, M.; Martindale, R.; Hurt, R.; Daley, B. Refeeding syndrome in the critically ill: A literature review and clinician's guide. *Curr. Gastroenterol. Rep.* **2019**, *21*, 58. [CrossRef]
4. Friedli, N.; Stanga, Z.; Sobotka, L.; Culkin, A.; Kondrup, J.; Laviano, A.; Mueller, B.; Schuetz, P. Revisiting the refeeding syndrome: Results of a systematic review. *Nutrition* **2017**, *35*, 151–160. [CrossRef] [PubMed]
5. Wirth, R.; Diekmann, R.; Janssen, G.; Fleiter, O.; Fricke, L.; Kreilkamp, A.; Modreker, M.K.; Marburger, C.; Nels, S.; Pourhassan, M.; et al. Refeeding syndrome: Pathophysiology, risk factors, prevention, and treatment. *Internist* **2018**, *59*, 326–333. [CrossRef] [PubMed]
6. Schuetz, P.; Fehr, R.; Baechli, V.; Geiser, M.; Deiss, M.; Gomes, F.; Kutz, A.; Tribolet, P.; Bregenzer, T.; Braun, N.; et al. Individualised nutritional support in medical inpatients at nutritional risk: A randomised clinical trial. *Lancet* **2019**, *393*, 2312–2321. [CrossRef] [PubMed]
7. Wunderle, C.; Siegenthaler, J.; Seres, D.; Owen-Michaane, M.; Tribolet, P.; Stanga, Z.; Mueller, B.; Schuetz, P. Adaptation of nutritional risk screening tools may better predict response to nutritional treatment: A secondary analysis of the randomized controlled trial Effect of early nutritional therapy on Frailty, Functional Outcomes, and Recovery of malnourished medical inpatients Trial (EFFORT). *Am. J. Clin. Nutr.* **2024**, *119*, 800–808. [PubMed]
8. Bioletto, F.; Pellegrini, M.; Ponzo, V.; Cioffi, I.; De Francesco, A.; Ghigo, E.; Bo, S. Impact of refeeding syndrome on short- and medium-term all-cause mortality: A systematic review and meta-analysis. *Am. J. Med.* **2021**, *134*, 1009–1018.e1. [CrossRef] [PubMed]
9. Crook, M.A.; Hally, V.; Panteli, J.V. The importance of the refeeding syndrome. *Nutrition* **2001**, *17*, 632–637. [CrossRef]
10. Crook, M.; Swaminathan, R. Disorders of plasma phosphate and indications for its measurement. *Ann. Clin. Biochem.* **1996**, *33*, 376–396. [CrossRef]
11. Valla, F.V.; Berthiller, J.; Gaillard-Le-Roux, B.; Ford-Chessel, C.; Ginhoux, T.; Rooze, S.; Cour-Andlauer, F.; Meyer, R.; Javouhey, E. Faltering growth in the critically ill child: Prevalence, risk factors, and impaired outcome. *Eur. J. Pediatr.* **2018**, *177*, 345–353. [CrossRef] [PubMed]
12. Hubert, A.; Ford-Chessel, C.; Berthiller, J.; Peretti, N.; Javouhey, E.; Valla, F.V. Nutritional status in pediatric intermediate care: Assessment at admission, progression during the stay and after discharge. *Arch. Pediatr.* **2016**, *23*, 333–339. [CrossRef] [PubMed]
13. Mehta, N.M.; Bechard, L.J.; Cahill, N.; Wang, M.; Day, A.; Duggan, C.P.; Heyland, D.K. Nutritional practices and their relationship to clinical outcomes in critically ill children—An international multicenter cohort study. *Crit. Care Med.* **2012**, *40*, 2204–2211. [CrossRef] [PubMed]
14. Mbethe, A.P.; Mda, S. Incidence of refeeding syndrome and its associated factors in South African children hospitalized with severe acute malnutrition. *Iran. J. Pediatr.* **2017**, *27*, e8297. [CrossRef]
15. Ross, J.R.; Finch, C.; Ebeling, M.; Taylor, S.N. Refeeding syndrome in very-low-birth-weight intrauterine growth-restricted neonates. *J. Perinatol.* **2013**, *33*, 717–720. [CrossRef] [PubMed]
16. Bonsante, F.; Iacobelli, S.; Latorre, G.; Rigo, J.; De Felice, C.; Robillard, P.Y.; Gouyon, J.B. Initial amino acid intake influences phosphorus and calcium homeostasis in preterm infants—It is time to change the composition of the early parenteral nutrition. *PLoS ONE* **2013**, *8*, e72880. [CrossRef] [PubMed]
17. Moltu, S.J.; Strømmen, K.; Blakstad, E.W.; Almaas, A.N.; Westerberg, A.C.; Brække, K.; Rønnestad, A.; Nakstad, B.; Berg, J.P.; Veierød, M.B.; et al. Enhanced feeding in very-low-birth-weight infants may cause electrolyte disturbances and septicemia—A randomized, controlled trial. *Clin. Nutr.* **2013**, *32*, 207–212. [CrossRef] [PubMed]
18. Boubred, F.; Herlenius, E.; Bartocci, M.; Jonsson, B.; Vanpée, M. Extremely preterm infants who are small for gestational age have a high risk of early hypophosphatemia and hypokalemia. *Acta Paediatr.* **2015**, *104*, 1077–1083. [CrossRef]
19. Chaudhary, V.; Gohil, J.R.; Patel, S.A. Hypophosphatemia in refeeding syndrome in intrauterine growth restricted IUGR neonates who are receiving nutrition: A prospective observational study. *Asian J. Pediatr. Res.* **2020**, *3*, 23–29. [CrossRef]
20. Cormack, B.E.; Jiang, Y.; Harding, J.E.; Crowther, C.A.; Bloomfield, F.H.; ProVIDe Trial Group. Neonatal refeeding syndrome and clinical outcome in extremely low-birth-weight babies: Secondary cohort analysis from the ProVIDe trial. *JPEN J. Parenter. Enter. Nutr.* **2021**, *45*, 65–78. [CrossRef]
21. Sung, S.I.; Chang, Y.S.; Choi, J.H.; Ho, Y.; Kim, J.; Ahn, S.Y.; Park, W.S. Increased risk of refeeding syndrome-like hypophosphatemia with high initial amino acid intake in small-for-gestational-age, extremely-low-birthweight infants. *PLoS ONE* **2019**, *14*, e0221042. [CrossRef] [PubMed]
22. Ichikawa, G.; Watabe, Y.; Suzumura, H.; Sairenchi, T.; Muto, T.; Arisaka, O. Hypophosphatemia in small for gestational age extremely low birth weight infants receiving parenteral nutrition in the first week after birth. *J. Pediatr. Endocrinol. Metab.* **2012**, *25*, 317–321. [CrossRef]
23. Al-Wassia, H.; Lyon, A.W.; Rose, S.M.; Sauve, R.S.; Fenton, T.R. Hypophosphatemia is prevalent among preterm infants less than 1500 grams. *Am. J. Perinatol.* **2019**, *36*, 1412–1419. [PubMed]
24. Senterre, T.; Abu Zahirah, I.; Pieltain, C.; de Halleux, V.; Rigo, J. Electrolyte and mineral homeostasis after optimizing early macronutrient intakes in VLBW infants on parenteral nutrition. *J. Pediatr. Gastroenterol. Nutr.* **2015**, *61*, 491–498. [CrossRef] [PubMed]

25. Bustos Lozano, G.; Hidalgo Romero, Á.; Melgar Bonis, A.; Ureta Velasco, N.; Orbea Gallardo, C.; Pallás Alonso, C. Early hypophosphataemia in at risk newborns. Frequency and magnitude. *Ann. Paediatr.* **2018**, *88*, 216–222. [CrossRef] [PubMed]
26. Brener Dik, P.H.; Galletti, M.F.; Fernández Jonusas, S.A.; Alonso, G.; Mariani, G.L.; Fustiñana, C.A. Early hypophosphatemia in preterm infants receiving aggressive parenteral nutrition. *J. Perinatol.* **2015**, *35*, 712–715. [CrossRef] [PubMed]
27. Bustos Lozano, G.; Soriano-Ramos, M.; Pinilla Martín, M.T.; Chumillas Calzada, S.; García Soria, C.E.; Pallás-Alonso, C.R. Early hypophosphatemia in high-risk preterm infants: Efficacy and safety of sodium glycerophosphate from first Day on parenteral nutrition. *JPEN J. Parenter. Enteral Nutr.* **2019**, *43*, 419–425. [CrossRef] [PubMed]
28. Rigo, J.; Pieltain, C.; Viellevoye, R.; Bagnoli, F. Calcium and phosphorus homeostasis: Pathophysiology. In *Neonatology*; Buonocore, G., Bracci, R., Weindling, M., Eds.; Springer: Milano, Italy, 2012; pp. 333–353.
29. Igarashi, A.; Okuno, T.; Ohta, G.; Tokuriki, S.; Ohshima, Y. Risk factors for the development of refeeding syndrome-like hypophosphatemia in very low birth weight infants. *Dis. Markers* **2017**, *2017*, 9748031. [CrossRef] [PubMed]
30. Pająk, A.; Królak-Olejnik, B.; Szafrańska, A. Early hypophosphatemia in very low birth weight preterm infants. *Adv. Clin. Exp. Med.* **2018**, *27*, 841–847. [CrossRef]
31. Mulla, S.; Stirling, S.; Cowey, S.; Close, R.; Pullan, S.; Howe, R.; Radbone, L.; Clarke, P. Severe hypercalcaemia and hypophosphataemia with an optimised preterm parenteral nutrition formulation in two epochs of differing phosphate supplementation. *Arch. Dis. Child. Fetal Neonatal Ed.* **2017**, *102*, F451–F455. [CrossRef]
32. Amissah, E.A.; Brown, J.; Harding, J.E. Protein supplementation of human milk for promoting growth in preterm infants. *Cochrane Database Syst. Rev.* **2018**, *6*, CD000433. [CrossRef] [PubMed]
33. Bellagamba, M.P.; Carmenati, E.; D'Ascenzo, R.; Malatesta, M.; Spagnoli, C.; Biagetti, C.; Burattini, I.; Carnielli, V.P. One extra gram of protein to preterm infants from birth to 1800 g: A single-blinded randomized clinical trial. *J. Pediatr. Gastroenterol. Nutr.* **2016**, *62*, 879–884. [CrossRef] [PubMed]

nutrients

MDPI

Article

Introduction of Solid Foods in Preterm Infants and Its Impact on Growth in the First Year of Life—A Prospective Observational Study

Margarita Thanhaeuser [1], Melanie Gsoellpointner [2], Margit Kornsteiner-Krenn [1], David Steyrl [3], Sophia Brandstetter [1], Bernd Jilma [2], Angelika Berger [1] and Nadja Haiden [4,*]

1 Department of Pediatrics and Adolescent Medicine, Comprehensive Center for Pediatrics, Medical University of Vienna, 1090 Vienna, Austria; margarita.thanhaeuser@meduniwien.ac.at (M.T.); margit.kornsteiner.krenn@gmail.com (M.K.-K.); sophia.brandstetter@meduniwien.ac.at (S.B.); angelika.berger@meduniwien.ac.at (A.B.)

2 Department of Clinical Pharmacology, Medical University of Vienna, 1090 Vienna, Austria; melanie.gsoellpointner@meduniwien.ac.at (M.G.); bernd.jilma@meduniwien.ac.at (B.J.)

3 Department of Cognition, Emotion, and Methods in Psychology, Faculty of Psychology, University of Vienna, 1010 Vienna, Austria; david.steyrl@univie.ac.at

4 Department of Neonatology, Kepler University Hospital, 4020 Linz, Austria

* Correspondence: nadja.haiden@kepleruniklinikum.at

Abstract: The aim of this study was to investigate whether age at introduction of solid foods in preterm infants influences growth in the first year of life. This was a prospective observational study in very low birth weight infants stratified to an early (<17 weeks corrected age) or a late (≥17 weeks corrected age) feeding group according to the individual timing of weaning. In total, 115 infants were assigned to the early group, and 82 were assigned to the late group. Mean birth weight and gestational age were comparable between groups (early: 926 g, 26 + 6 weeks; late: 881 g, 26 + 5 weeks). Mean age at weaning was 13.2 weeks corrected age in the early group and 20.4 weeks corrected age in the late group. At 12 months corrected age, anthropometric parameters showed no significant differences between groups (early vs. late, mean length 75.0 vs. 74.1 cm, weight 9.2 vs. 8.9 kg, head circumference 45.5 vs. 45.0 cm). A machine learning model showed no effect of age at weaning on length and length z-scores at 12 months corrected age. Infants with comorbidities had significantly lower anthropometric z-scores compared to infants without comorbidities. Therefore, regardless of growth considerations, we recommend weaning preterm infants according to their neurological abilities.

Keywords: preterm infants; solid foods; growth; necrotizing enterocolitis; bronchopulmonary dysplasia; intraventricular hemorrhage; machine learning

1. Introduction

From around 6 months of age, the nutritional needs of infants can no longer be met by human milk or formula [1]. Therefore, the European Society for Pediatric Gastroenterology, Hepatology, and Nutrition (ESPGHAN) recommends the introduction of complementary foods between the 17th and the 26th week of life in term infants, but evidence-based guidelines for the introduction of solid foods in preterm infants are not yet available [2]. Guidelines on the optimal composition of complementary foods for preterm infants are also lacking [2].

The increased nutritional requirements of preterm infants after birth and following their discharge from hospital differ significantly from those of full-term infants [3]. This suggests that special nutritional requirements may also be needed during weaning to achieve adequate growth in preterm infants.

Numerous observational studies on weaning in preterm infants have been published, showing a trend towards very early initiation of complementary feeding in these infants [4–8]. The mean age at weaning was earlier than the recommended age for term infants in almost all published observational studies, with the youngest infants starting earliest. However, the associated growth outcomes have shown considerable variability [4–8] and introduction of solid foods before 4 months of age might be associated with an increased risk of allergy [2].

Only three randomized controlled trials on the introduction of solid foods in preterm infants have been published, one of them more than 20 years ago and one in an emerging country [9,10]. Thus, both studies cannot be compared to current or Western standards. The third study was published recently and included only infants with stable growth [11]. Data on infants with comorbidities due to preterm birth are lacking.

The primary aim of this observational study was to investigate whether the timing of the introduction of solid foods in preterm infants has an impact on growth in the first year of life. Our goal was to encompass infants with significant perinatal conditions that could potentially impact their stable growth as described in the literature, such as necrotizing enterocolitis (NEC) [12] and chronic lung disease (bronchopulmonary dysplasia—BPD) [13] or intraventricular hemorrhage (IVH) due to negative influences on suck–swallow rhythms [14]. We aimed to closely monitor the introduction of solid foods in these infants and assess the advantages of introducing solid foods at different ages and with varying compositions of complementary foods.

2. Materials and Methods

This was a prospective observational study in preterm infants with a birth weight < 1500 g and a gestational age < 32 weeks. Infants were recruited between April 2016 and November 2021 in the neonatal outpatient clinic of a level IV neonatal care unit of the Medical University of Vienna, Comprehensive Center for Pediatrics, Austria, after informed consent was obtained from the parents. Written informed consent from one parent was sufficient due to low risk for participants. The study was approved by the ethics committee of the Medical University of Vienna (EK: 1273/2016, date of approval 16 April 2016) and registered at clinicaltrials.gov (NCT02936219, 18 October 2016).

At enrollment, infants were stratified according to their type of milk feedings (human milk, formula milk, or mixed feedings). Based on the specific time chosen by each parent for the introduction of solid foods, we categorized infants into two groups: an early complementary feeding group (starting solid foods <17th week of life corrected age) and a late complementary feeding group (starting solid foods ≥17th week of life corrected age).

2.1. Study Participants

All infants with a birth weight < 1500 g and a gestational age < 32 + 0 visiting the neonatal outpatient clinic of the Medical University of Vienna were eligible for the study. Infants with gastrointestinal diseases such as Hirschsprung's disease, congenital heart disease, major congenital birth defects, or chromosomal aberrations were excluded from the trial.

2.2. Study Visits

Families of participating infants were invited to study visits together with regular visits at the neonatal outpatient clinic at term, 6 weeks, 12 weeks, 6 months, 9 months, and 12 months corrected age. For the study flow, please refer to Figure S1, Supplemental Material.

Anthropometry was assessed under standardized conditions at every visit using a baby scale (Seca 376, Seca Germany, Hamburg, Germany) in lying position for weight. Length was measured in lying position (Seca 210, Seca Germany, Hamburg, Germany). Head circumference measurements were performed with a tape measure.

Nutritional intake was estimated based on 24 h recalls at term and at 6 weeks corrected age. Furthermore, three-day dietary records (a self-reported logbook with a food record on three consecutive days, including one weekend day) and the introduction of the main food categories were queried at 3 months, 6 months, 9 months, and 12 months corrected age.

Data on comorbidities of infants were retrieved from medical charts.

2.3. Primary Outcome

With this observational study, we aimed to identify current feeding practices in preterm infants after starting solid foods. The primary outcome of this study was to examine whether age at introduction of solid foods had an influence on length at 12 months corrected age. To detect a difference in length of 5% between study groups, the inclusion of 152 infants was necessary.

2.4. Secondary Outcomes

Secondary outcomes included other anthropometric parameters and their corresponding z-scores. Influences of different comorbidities (NEC $\geq$ grade 2, BPD defined as oxygen demand $\geq 36 + 0$, retinopathy of prematurity—ROP $\geq$ grade 2, culture-proven sepsis, IVH $\geq$ grade 2) on the age at introduction of solid foods were assessed.

2.5. Baseline Characteristics

Maternal and infant baseline characteristics, as well as data on neonatal morbidities, were collected from medical charts. Data on parental education were collected in the follow-up visits and divided into three groups according to the highest education level of either the father or mother of the infant (primary, secondary, or tertiary school).

2.6. Statistical Analysis and Machine Learning Model

In general, absolute and relative frequencies were calculated for ordinal and nominal data obtained on any date of measurement, respectively. For continuous variables, mean and standard deviations were calculated, and graphical descriptive analyses included growth curves and a dependency plot. Dependencies between siblings of multiple birth were not considered in the descriptive analysis. To detect differences between study groups, the Chi2-test or the Mann–Whitney U-test was applied. For basic statistical analyses, the program JASP version 0.18.3 was used.

For the regression analysis of the primary outcome length and length z-score at 12 months corrected age, a machine learning model was fitted. Furthermore, another model was fitted to detect the most influential parameters on the timepoint of starting solid foods to predict age at weaning. A total of five machine learning-based statistical analyses were performed using Python version 3.11 [15]. There were two to predict length at 12 months corrected age with a continuous and a categorical variant of the variable age at introduction of solid foods, and there were two to predict length z-score at 12 months corrected age, again with a continuous and a categorical variant of the variable age at introduction of solid foods. For prediction of length and length z-score at 12 months corrected age, the following confounding variables were included: gestational age, sex, length z-score at term, type of feeding at 6 weeks corrected age, age at introduction of solid foods, height of mother and father, BPD, and NEC.

The fifth model was performed to predict age when starting solid foods and included the following confounding variables: gestational age, sex, type of feeding at 6 weeks corrected age, BPD, NEC, IVH, maternal country of birth, maternal age, and highest parental education.

Each analysis includes the following three parts: (1) prediction model training [16,17], (2) generalizability testing [18,19], (3) model analysis [20,21]. Exact details on the analyses can be found in Appendix A.

3. Results

3.1. Screening and Participants

During the 5.5 year study period between April 2016 and November 2021, 580 infants were screened. A total of 529 infants met the inclusion criteria; in 308 cases, the parents refused participation, and 3 infants had already started solid foods at their first appointment in the neonatal outpatient clinic. After 21 dropouts due to various reasons (i.e., withdrawal of consent, screening failure, lost to follow up, no data on starting solid foods or on the primary outcome), the final cohort consisted of 197 infants. According to their corrected age when starting solid foods, 115 infants were assigned to the early group and 82 to the late group (Figure 1).

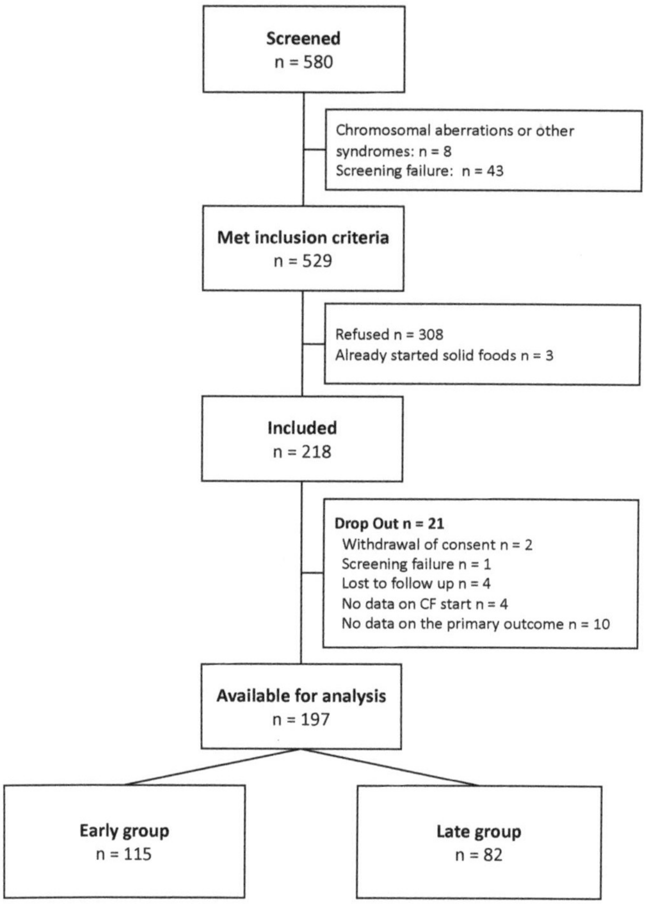

Figure 1. Patient flow chart.

3.2. Baseline Characteristics and Neonatal Morbidity

Table 1 shows the maternal and infant baseline characteristics, as well as data on neonatal morbidities.

Table 1. Baseline characteristics and neonatal morbidity.

Parameter	Early Group (n = 115)	Late Group (n = 82)
Obstetric and parental parameters		
Multiple pregnancy	36 (31.3%)	19 (23.2%)
Cesarean delivery	95 (82.6%)	70 (85.4%)
Prenatal steroids (any)	105 (91.3%)	76 (92.7%)
Premature rupture of membranes	43 (37.4%)	42 (51.2%) *
Gestational diabetes	0 (0%)	3 (3.7%) *
Preeclampsia	13 (11.3%)	17 (20.7%)
Age of mother at birth	31.4 (±5.8)	33.2 (±5.3)
Age of father at birth	35.1 (±7.2)	35.2 (±6.5)
Maternal education		
No graduation/school diploma	12 (10.4%)	8 (9.8%)
Middle school	32 (27.8%)	19 (23.2%)
Secondary school	23 (20%)	16 (19.5%)
Post-secondary school	43 (37.4%)	36 (43.9%)
Paternal education		
No graduation/school diploma	10 (8.7%)	8 (9.8%)
Middle school	45 (39.1%)	24 (29.3%)
Secondary school	21 (18.3%)	20 (24.4%)
Post-secondary school	33 (28.7%)	24 (29.3%)
Neonatal parameters		
Male sex	69 (60%)	36 (43.9%) *
Gestational age (days)	26 + 6 (±2 + 0)	26 + 5 (±2 + 2)
Birth weight (g)	926 (±254)	881 (±262)
Small for gestational age	4 (3.5%)	4 (4.9%)
Neonatal morbidity		
Necrotizing enterocolitis ≥ grade II	5 (4.3%)	6 (7.3%)
Bronchopulmonary dysplasia	14 (12.2%)	23 (28%) *
Persisting ductus arteriosus	51 (44.3%)	47 (57.3%)
Retinopathy of prematurity ≥ grade II	34 (29.6%)	27 (32.9%)
Sepsis, culture positive	16 (13.9%)	19 (23.2%)
Intraventricular hemorrhage ≥ grade II	17 (14.8%)	12 (14.6%)
Periventricular leukomalacia	0 (0%)	1 (1.2%)

Categorical data are presented as numbers with percentages in parentheses. Continuous data are presented as the mean and standard deviation in parentheses. * marks significant difference.

Mothers of infants in the late feeding group had a significantly higher incidence of premature rupture of membranes ($p = 0.049$) and gestational diabetes ($p = 0.038$).

Infants in the early group started solid foods at a mean age of 13.2 weeks corrected age (CI 95% 12.7–13.8, ±3), while infants in the late group started at 20.4 weeks corrected age (CI 95% 19.8–21.0, ±2.9).

The percentage of male infants was significantly higher in the early feeding group ($p = 0.026$), whereas BPD rate was significantly higher in the late feeding group ($p = 0.005$). With a mean gestational age of 26 + 6 (CI 95% 26 + 3–27 + 2, ±2 + 0) compared to 26 + 5 (CI 95% 26 + 2–27 + 2, ±2 + 2), infants of the early feeding group were of similar gestational age but had a slightly higher birth weight (early group: mean 926 g, CI 95% 879–973, ±254, late group: mean 881 g, CI 95% 823–938, ±262, p = n.s.). Other parameters were comparable between study groups.

3.3. Primary Outcome

At 12 months corrected age, infants in the early group had a mean length of 75.0 cm (CI 95% 74.4–75.5, ±3.1); infants in the late group were 0.9 cm shorter with a mean length of 74.1 cm (CI 95% 73.4–74.8, ±3.3) (Table S1, Supplemental Material). This difference was insignificant ($p = 0.053$). Figure 2A depicts length z-scores within the first year of life.

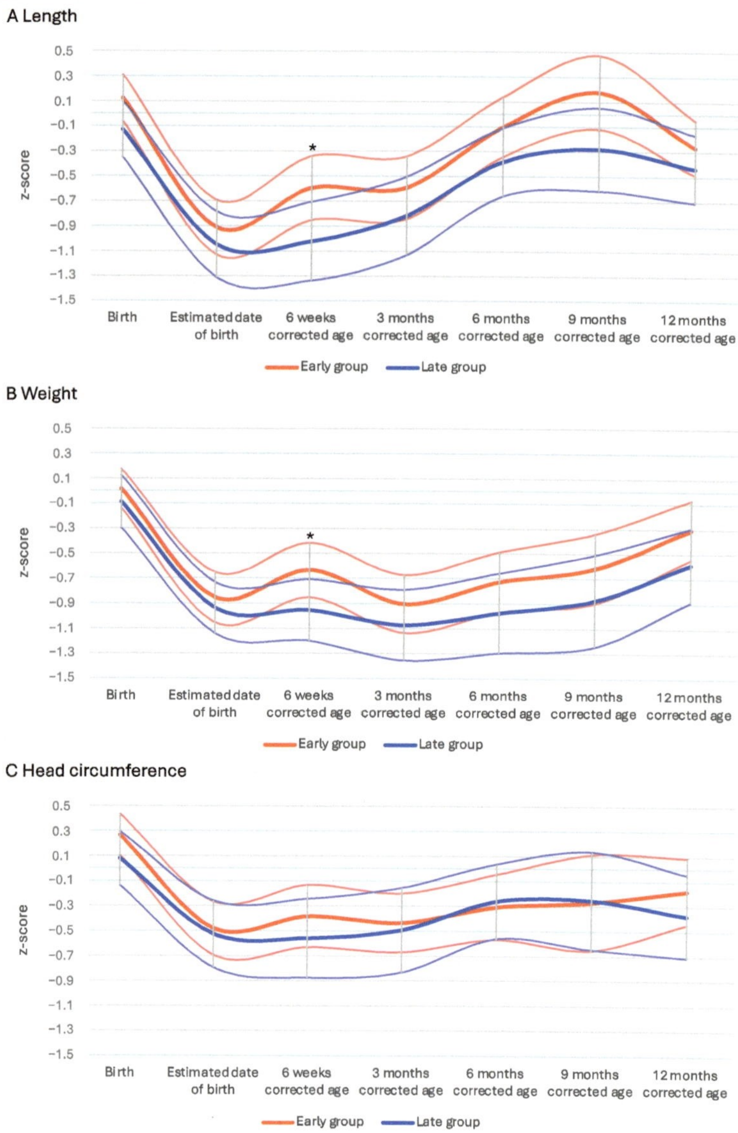

Figure 2. Z-scores of anthropometric data of the study population. Plot (**A**): this plot shows the length z-scores (mean and CI 95%) in the first year of life. The asterisks mark a significant *p*-value < 0.05. Plot (**B**): this plot shows the weight z-scores (mean and CI 95%) in the first year of life. The asterisks mark a significant *p*-value < 0.05. Plot (**C**): this plot shows the head circumference z-scores (mean and CI 95%) in the first year of life.

3.4. Secondary Outcomes

Table S1, Supplemental Material, shows all anthropometric parameters assessed in the first year of life of infants, including corresponding z-scores. Figure 2B,C depict weight and head circumference within the first year of life. At 6 weeks corrected age, the late feeding group exhibited significantly lower z-scores for both weight and length compared to the early feeding group. Specifically, the weight z-score mean for the early group was −0.64 (CI 95% −0.85–−0.42, ±1.15), while for the late group, it was −0.96 (CI 95% −1.20–−0.71,

±1.08) (p = 0.043). For length, the early group had a mean z-score of −0.597 (CI 95% −0.85–−0.34, ±1.34), and the late group had −1.02 (CI 95% −1.34–−0.71, ±1.39) (p = 0.036). However, these differences occurred before the introduction of solid foods in both groups. At 3 and 6 months corrected age, length of infants was higher in the early feeding group (3 months corrected age—early group: mean 59.8 cm (CI 95% 59.3–60.4, ±2.8), late group: mean 59.1 cm (CI 95% 58.4–59.8, ±2.9), p = 0.041; 6 months corrected age—early group: mean 66.9 cm (CI 95% 66.4–67.5, ±2.9), late group: mean 65.8 cm (CI 95% 65.2–66.5, ±2.9), p = 0.008). However, this difference in length did not persist up to 12 months corrected age.

3.5. *Influence of Comorbidities, Type of Feeding, and Birthweight on Introduction of Solids*

Table 2 presents the corrected age in weeks at which solid foods were introduced to infants with various comorbidities. It was observed that among infants with different neonatal conditions, those diagnosed with BPD started weaning the latest, at a mean corrected age of 18.1 weeks. Additionally, the table illustrates variations in the age at weaning based on the type of feeding at 6 weeks corrected age and according to birth weight.

Table 2. Weeks corrected age when starting solid foods in infants according to different comorbidities, type of feeding at 6 weeks corrected age, and birth weight.

		Total	Early Group	Late Group
Morbidities	NEC ≥ grade II (n = 11)	17.5 (±2.2)	15.8 (±1.6)	19 (±1.6)
	BPD (n = 37)	18.1 (±4.5)	14.1 (±3.2)	20.6 (±3.3)
	ROP ≥ grade II (n = 61)	16.9 (±4.1)	14.1 (±2.6)	20.4 (±2.7)
	Sepsis, culture positive (n = 35)	16.9 (±3.8)	13.6 (±2.3)	19.7 (±2.2)
	IVH ≥ grade II (n = 29)	16.9 (±4.2)	14.2 (±2.5)	20.8 (±2.9)
Milk	Breast milk (n = 62)	17.6 (±4.3)	13.8 (±1.7)	20.6 (±3.1)
	Mixed feedings (n = 33)	16.1 (±4.9)	12.4 (±3.1)	20.5 (±2.6)
	Formula (n = 97)	15.5 (±3.9)	13.5 (±3.1)	19.6 (±1.8)
Weight	<750 g (n = 62)	16.8 (±4.6)	13.7 (±3.2)	20.5 (±3.0)
	750–1000 g (n = 57)	16.0 (±4.0)	13.5 (±2.5)	20.1 (±2.3)
	>1000 g (n = 77)	15.9 (±5.0)	12.8 (±3.2)	20.5 (±3.2)

Continuous data are presented as the mean weeks corrected age with standard deviation in parentheses. BPD—bronchopulmonary dysplasia defined as oxygen demand > 36 + 0, IVH—intraventricular hemorrhage, NEC—necrotizing enterocolitis, ROP—retinopathy of prematurity.

Overall, breastfeeding was more prevalent in the late group compared to the early group. Specifically, 28% of infants in the early group were exclusively breastfed, while 65% were formula-fed. In contrast, 34% of infants in the late group were exclusively breastfed, and 31% were on formula feeding. Additionally, infants who were exclusively breastfed began solid foods approximately two weeks later than those fed formula milk. Within our study cohort, the age at which weaning occurred was inversely related to birth weight, indicating that infants with lower birth weights started solid foods later.

Data on anthropometric z-scores in infants with different comorbidities are shown in Table S2, Supplemental Material. Infants with NEC showed an especially good catch-up growth after starting solid foods.

3.6. *Machine Learning Models*

Table 3 shows the results of the four machine learning models used to predict the primary outcome length and length z-score at 12 months corrected age. Age at introduction of solid foods neither influenced length at 12 months corrected age nor its z-score. Also, there was no difference whether using the categorical variable (early vs. late feeding group) or the continuous variable (age at introduction of solid foods in weeks). Length at term, sex, and height of the mother showed a significant influence on the primary outcome length at 12 months of age, whereas the model fit was low for all four models.

Table 3. Machine learning model, including influential factors for the prediction of length and length z-score at 12 months corrected age.

	Length at 12 Months Corrected Age				**Length z-Score at 12 Months Corrected Age**			
	Early group vs. Late group		Weeks corrected age at starting solids		Early group vs. Late group		Weeks corrected age at starting solids	
Model fit	$R^2 = 0.138$		$R^2 = 0.134$		$R^2 = 0.134$		$R^2 = 0.125$	
	Effect size	*p-value*	*Effect size*	*p-value*	*Effect size*	*p-value*	*Effect size*	*p-value*
Length z-score at term	1.03	**<0.001**	0.99	**<0.001**	0.39	**<0.001**	0.39	**<0.001**
Female sex	0.48	**0.001**	0.49	**<0.001**	0.09	0.116	0.09	0.157
Height of mother	0.3	**0.039**	0.26	**0.039**	0.11	**0.015**	0.1	0.066
Age at introduction of solids	0.19	0.181	0.14	0.560	0.04	0.542	0.03	0.843
Nutrition at 6 weeks	0.14	0.633	0.11	0.912	0.05	0.719	0.05	0.680
BPD	0.08	0.549	0.11	0.278	0.02	0.675	0.03	0.541
Height of father	0.06	0.939	0.09	0.922	0.03	0.917	0.04	0.942
Gestational age	0.05	0.885	0.07	0.858	0.02	0.921	0.03	0.916
NEC	0.03	0.915	0.01	0.894	0.01	0.905	0	0.866

For the columns "Early vs. Late group", the variable "age at introduction of solid foods" was categorical with a cut-off at 17th weeks corrected age; for the columns "Weeks corrected age at starting", the variable "age at introduction of solid foods" was continuous. Significant *p*-values < 0.5 are bold. BPD—bronchopulmonary disease with oxygen demand > 36 + 0 weeks corrected age, NEC—necrotizing enterocolitis.

The dependency plot in Figure S2, Supplemental Material shows the effect of age when starting solid foods on length at 12 months corrected age. Between 15 and 18 weeks corrected age, there is a turning point—starting on solid foods before has a slightly positive effect on length at 12 months corrected age; starting thereafter has a negative effect.

Another model for the prediction of age when starting solid foods was used (Table S3, Supplemental Material), but other than BPD, no significant influential factors could be detected. Again, the model fit was very low.

4. Discussion

This study is a prospective observational analysis of preterm infants, exploring how the timing of introducing complementary foods affects length and other anthropometric measures during the first year of life. The introduction of solid foods had no impact on growth within the first year of life for infants with an average birth weight below 1000 g and a mean gestational age at birth of less than 28 weeks in both groups. However, at 3 and 6 months corrected age, infants in the early feeding group exhibited greater length, suggesting a temporary acceleration in length growth. Nevertheless, other anthropometric measurements showed no significant differences between the groups during the first year. Infants in both study groups showed a remarkable length and weight catch-up growth from birth until one year corrected age.

Our findings align with the majority of both interventional and observational studies on the introduction of solid foods in preterm infants and their impact on growth. Only a few studies have reported an effect of early introduction of solid foods on subsequent growth. An observational study by Brion et al. included infants with a gestational age < 28 weeks and compared infants receiving ready-made complementary foods with those receiving home-made complementary foods [22]. Infants were grouped according to the age at introduction of solid foods. For infants on ready-made complementary foods starting at <26 weeks corrected age, the z-scores for weight-for-length and BMI were highest at one year corrected age [22].

Another study that reported higher growth velocity is a randomized controlled trial by Marriott et al., published over 20 years ago [9]. In this study, infants were assigned to either a group following a preterm weaning strategy or a control group adhering to the current best practice at that time. Infants in the preterm weaning strategy group were started on solid foods at 13 weeks postnatal age and received complementary foods that were higher in energy density and protein content. Consequently, it is not surprising that these infants had a higher growth rate compared to those in the control group [9].

Additionally, two other interventional randomized controlled trials by Gupta et al. and Haiden et al. have explored early versus late introduction of complementary feeding in preterm infants [10,11]. Haiden et al. observed a transient faster weight gain in the early group, with a higher weight-for-age z-score at 6 months corrected age. However, both studies found no differences in anthropometric parameters, such as weight, length, or head circumference, at one year corrected age [11].

Many observational studies have also reported no impact of the timing of solid food introduction on growth [8,23,24]. Therefore, it can be concluded that the introduction of solid foods to preterm infants should not be solely based on growth considerations, a finding that is further supported by our results [11]. The timing of weaning should instead be determined by the infants' neurological abilities and readiness.

Almost all observational studies indicate that preterm infants are being introduced to solid foods earlier than recommended for term infants, with the earliest weaning occurring in the most preterm-born infants. One study from the United States reported that preterm infants were more likely to be introduced to solid foods before 4 months corrected age [4]. Similar findings emerged from Australia, where the median time for starting solid foods was 14 weeks corrected age, whereas term infants were weaned 5 weeks later [5]. In a study from Italy, median age to start solids was 15 weeks corrected age, with 18% of infants weighing less than <5 kg at weaning [6]. An Austrian study by Hofstaetter et al. revealed that more than 50% of infants were on solid foods before 17 weeks corrected age, with 23% starting before 12 weeks [7].

In our cohort, the mean age of starting solid foods was 16.2 weeks, with almost 60% of infants starting before 17 weeks corrected age. The mean age of weaning was 13.2 weeks in the early group and 20.4 weeks in the late group. Infants in the late group had a lower birth weight and a lower z-score for weight and length at 6 weeks corrected age. Unlike the findings of the above-mentioned studies, birth weight in our cohort inversely correlated with age at weaning. We did not inquire about the reasons why parents chose to introduce solid foods early. This would have provided some insight. It is presumed that many followed their pediatrician's advice, as was also noted in a study by Baldassarre et al. [25]. The authors highlighted significant variability in weaning advice from primary care pediatricians due to the absence of evidence-based guidelines [25].

To identify the factors that most influence the age at which solid foods are introduced, we used a machine learning model to predict age at weaning, incorporating a range of variables, including infant baseline characteristics, maternal factors, neonatal morbidity, and parental cultural and socioeconomic background. Despite the broad scope of factors considered, the model demonstrated poor predictive performance, suggesting key influencing factors were not included in our analysis. Notably, as mentioned above, we collected neither information on parents' reasons for introducing solid foods nor data on the infants' neurological abilities at the time of introducing complementary feeding, both of which could potentially have improved predictability. Among variables we did examine, only BPD showed a significant relationship with weaning age, indicating that infants with BPD were more likely to start solid foods later, possibly because of ongoing respiratory instability or problems in oromotor function after long-term respiratory support. Other factors such as sex, gestational age, type of nutrition at six weeks corrected age, IVH, NEC, parental educational levels, and maternal age or country of birth did not significantly influence age at introduction of solids foods in our cohort. Although we observed a significantly higher rate of breastfeeding at six weeks corrected age in the late feeding group, breastfeeding was not found to be an influential factor in predicting age at weaning.

4.1. Comorbidities

Percentages of infants with a birth weight < 1000 g were rather low in other observational studies on preterm infants and solid foods, except for the studies by Spiegler et al.,

Ribas et al., and Boscarino et al., which focused on VLBW infants, and Brion et al., which included only infants with a birthweight < 1000 g [8,22,23,26]. To the best of our knowledge, none of these studies specifically targeted infants with significant comorbidities related to preterm birth. In our study, the incidence of comorbidities was comparable between the two study groups, except for the rate of BPD. However, we observed a tendency for parents of infants diagnosed with NEC, BPD, and sepsis to introduce solid foods later than parents of infants without such comorbidities. Infants with NEC especially showed a good catch-up growth after starting solid foods.

4.2. Limitations and Strengths

In 2022, we published a randomized controlled trial focusing on the introduction of solid foods in preterm infants [11]. To ensure a uniform cohort, we excluded infants with conditions that could affect stable growth. Noticing the lack of research on post-discharge nutritional interventions in infants with comorbidities, we conducted this prospective observational study. A limitation of our study is the lack of data on the neurological readiness of infants for solid foods and the lack of information on the reasons behind parents' decisions to introduce solid foods, whether due to the child's interest, pediatrician recommendations, or other factors. This information would have significantly enriched our understanding.

To address the inherent limitations of our study's design, we chose a powerful statistical approach using machine learning. This method focuses on prediction, i.e., out-of-sample inference using flexible, complex, high-dimensional models, that is both robust and credible rather than in-sample inference, as is typically found in statistical testing. We developed a prediction model for the primary outcome of length at 12 months corrected age and the corresponding z-scores, including variables such as the age at introduction of solid foods, nutrition at 6 weeks, sex, gestational age, length at term, height of mother and father, BPD, and NEC. However, the model fit was low, indicating that other unknown factors influencing length at 12 months corrected age were not captured.

Our study provides valuable data on the initiation of complementary feeding in a sizeable cohort of infants with a mean birth weight of <1000 g, including those with neonatal comorbidities. We have also assessed the nutritional data through monthly feeding protocols, although analyses of these findings are pending.

5. Conclusions

The timing of the introduction of solid foods did not affect growth in the first year of life in VLBW infants, regardless of the presence or absence of comorbidities. Preterm infants diagnosed with BPD and those who were breastfed at 6 weeks corrected age started solid foods the latest. Furthermore, there was an inverse relationship between birth weight and corrected age at the start of solid foods in our cohort.

Based on these findings, we conclude that the introduction of solid foods in preterm infants should be guided by the neurological abilities of the infants rather than their growth metrics or any neonatal morbidities.

Supplementary Materials: The following supporting information can be downloaded at https://www.mdpi.com/article/10.3390/nu16132077/s1. Figure S1: Study flow; Figure S2: Dependency plot—effect size of weeks corrected age when starting solid food on length at 12 months corrected age; Table S1: Anthropometric measurements; Table S2: Weight, length, and head circumference z-score in infants with comorbidities; Table S3: Machine learning model including influential factors for the prediction of age when starting solid foods.

Author Contributions: Conceptualization, N.H.; methodology, N.H., B.J., and M.T.; formal analysis, D.S., M.T., and N.H.; investigation, N.H.; resources, N.H. and A.B.; data curation, M.T., M.G., and M.K.-K.; writing—original draft preparation, M.T.; writing—review and editing, N.H., B.J., D.S., M.K.-K., M.G., S.B., and A.B.; visualization, M.T. and D.S.; supervision, N.H.; project administration, N.H.; funding acquisition, N.H. All authors have read and agreed to the published version of the manuscript.

Funding: The study was funded by the Buergermeisterfonds der Stadt Wien (medical-scientific fund of the mayor of Vienna, project number 15235) and by the Novalac International Pediatric Research Award (March 2016). Supported by Johannes Kepler University Open Access Publishing Fund.

Institutional Review Board Statement: The study was conducted in accordance with the Declaration of Helsinki and approved by the Institutional Review Board of the Medical University of Vienna (EK: 1273/2016, 16 April 2016) and registered on clinicaltrials.gov (NCT02936219, 18 October 2016).

Informed Consent Statement: Written informed consent from one parent was sufficient due to low risk for the participants.

Data Availability Statement: The study protocol and the individual participant data that underlie the results reported in this article, after de-identification, are available upon request from the corresponding author 6 months after publication. Researchers will need to state the aims of any analyses and provide a methodologically sound proposal. Proposals should be directed to nadja.haiden@meduniwien.ac.at. Data requestors will need to sign a data access agreement and, in keeping with patient consent for secondary use, obtain ethical approval for any new analyses due to ethical reasons.

Conflicts of Interest: Nadja Haiden reports consulting fees from Medis, MAM, Baxter, and Nestle and honoraria for lectures from Nestle, Baxter, Danone, and Hipp outside the submitted work. All other authors have no conflicts of interest to report.

Appendix A. Machine Learning Analysis

For the regression analysis of the primary outcome length and length z-score at 12 months corrected age, a machine learning model was fitted. Furthermore, another model was fit to detect the most influential parameters on the age when starting solid foods to predict age at weaning.

Therefore, a total of five machine learning-based statistical analyses were performed [15]: two to predict length at 12 months corrected age with a continuous and a categorical variant of the variable age at introduction of solid foods; two to predict length z-score at 12 months corrected age, again with a continuous and a categorical variant of the variable age at introduction of solid foods; and one more to predict age at starting solid foods.

Each analysis includes the following three parts: (1) prediction model training, (2) generalizability testing, and (3) model analysis.

(1) In the execution of the learning task, Gradient Boosted Decision Tree (GBDT) models were selected due to their demonstrated computational efficiency and high accuracy, as substantiated by Grinsztajn et al. and Ke et al. [16,17]. These models possess an intrinsic capability to capture non-linear associations and variable interactions [17]. Moreover, GBDT models exhibit robustness against multicollinearity and outliers [17].

(2) In order to evaluate the regression performance, particularly focusing on the generalizability and out-of-sample prediction accuracy, a nested cross-validation (CV) procedure was employed. CV is designed to provide a more robust assessment of the model's predictive capabilities beyond the confines of the training dataset, thereby offering a comprehensive insight into its real-world applicability and reliability [18]. CV implements repeated splits of the data into training and testing sets, whereas a 10 times 5-fold scheme is applied in the main (outer) CV loop. In each repetition of the main CV loop, the respective training set is used for data scaling (standardization) and model complexity tuning. Model complexity tuning is carried out in a nested (inner) CV procedure (10 times 5-fold) using a random search scheme. The complexity parameters that lead to the highest prediction accuracy in the inner CV procedure are subsequently used to train a GBDT model in the main CV loop. The model is subsequently tested on the respective testing set of the main CV loop. Regression performance is measured with the prediction coefficient of determination. Notably, the prediction R^2 will be smaller than R^2 values of conventional statistical models because the prediction R^2 measures prediction performance for unknown data and not post hoc model fit [19].

(3) The importance of single predictors for the model's performance was assessed with SHAP (Shapely Additive explanations) [20,21]. Originating from the domain of interpretable machine learning, SHAP leverages the concept of Shapley values from cooperative game theory. This approach quantitatively ascertains the impact of each predictor, including interaction effects, on the model's performance. The SHAP method is instrumental in discerning how individual predictors influence the model's predictions. By aggregating these contributions across numerous predictions, SHAP facilitates a thorough examination of the pivotal roles played by individual predictors in the context of the predictive task [20,21].

References

1. EFSA Panel on Nutrition, Novel Foods and Food Allergens (NDA); Castenmiller, J.; de Henauw, S.; Hirsch-Ernst, K.I.; Kearney, J.; Knutsen, H.K.; Maciuk, A.; Mangelsdorf, I.; McArdle, H.J.; Naska, A.; et al. Appropriate age range for introduction of complementary feeding into an infant's diet. *EFSA J.* **2019**, *17*, e05780. [CrossRef] [PubMed]
2. Fewtrell, M.; Bronsky, J.; Campoy, C.; Domellöf, M.; Embleton, N.; Fidler Mis, N.; Hojsak, I.; Hulst, J.M.; Indrio, F.; Lapillonne, A.; et al. Complementary Feeding: A Position Paper by the European Society for Paediatric Gastroenterology, Hepatology, and Nutrition (ESPGHAN) Committee on Nutrition. *J. Pediatr. Gastroenterol. Nutr.* **2017**, *64*, 119–132. [CrossRef] [PubMed]
3. Koletzko, B.; Cheah, F.C.; Domellof, M.; van Goudoever, J.B.; Poindexter, B.B.; Vain, N. Scientific Basis and Practical Application of Nutritional Care for Preterm Infants. *World Rev. Nutr. Diet.* **2021**, *122*, XIII–XIV. [CrossRef]
4. Braid, S.; Harvey, E.M.; Bernstein, J.; Matoba, N. Early introduction of complementary foods in preterm infants. *J. Pediatr. Gastroenterol. Nutr.* **2015**, *60*, 811–818. [CrossRef] [PubMed]
5. Cleary, J.; Dalton, S.M.; Harman, A.; Wright, I.M. Current practice in the introduction of solid foods for preterm infants. *Public Health Nutr.* **2020**, *23*, 94–101. [CrossRef]
6. Fanaro, S.; Borsari, G.; Vigi, V. Complementary feeding practices in preterm infants: An observational study in a cohort of Italian infants. *J. Pediatr. Gastroenterol. Nutr.* **2007**, *45* (Suppl. S3), S210–S214. [CrossRef]
7. Hofstatter, E.; Kottstorfer, V.; Stroicz, P.; Schutz, S.; Auer-Hackenberg, L.; Brandner, J.; Wald, M. Introduction and feeding practices of solid food in preterm infants born in Salzburg! *BMC Pediatr.* **2021**, *21*, 56. [CrossRef]
8. Boscarino, G.; Conti, M.G.; Pagano, F.; Di Chiara, M.; Pannucci, C.; Onesta, E.; Prota, R.; Deli, G.; Dito, L.; Regoli, D.; et al. Complementary Feeding and Growth in Infants Born Preterm: A 12 Months Follow-Up Study. *Children* **2021**, *8*, 1085. [CrossRef]
9. Marriott, L.D.; Foote, K.D.; Bishop, J.A.; Kimber, A.C.; Morgan, J.B. Weaning preterm infants: A randomised controlled trial. *Arch. Dis. Child. Fetal Neonatal Ed.* **2003**, *88*, F302–F307. [CrossRef]
10. Gupta, S.; Agarwal, R.; Aggarwal, K.C.; Chellani, H.; Duggal, A.; Arya, S.; Bhatia, S.; Sankar, M.J.; Sreenivas, V.; Jain, V.; et al. Complementary feeding at 4 versus 6 months of age for preterm infants born at less than 34 weeks of gestation: A randomised, open-label, multicentre trial. *Lancet Glob. Health* **2017**, *5*, e501–e511. [CrossRef]
11. Haiden, N.; Thanhaeuser, M.; Eibensteiner, F.; Huber-Dangl, M.; Gsoellpointner, M.; Ristl, R.; Kroyer, B.; Brandstetter, S.; Kornsteiner-Krenn, M.; Binder, C.; et al. Randomized Controlled Trial of Two Timepoints for Introduction of Standardized Complementary Food in Preterm Infants. *Nutrients* **2022**, *14*, 697. [CrossRef] [PubMed]
12. Bell, M.; Cole, C.R.; Hansen, N.I.; Duncan, A.F.; Hintz, S.R.; Adams-Chapman, I.; Eunice Kennedy Shriver National Institute of Child Health and Human Development Neonatal Research Network. Neurodevelopmental and Growth Outcomes of Extremely Preterm Infants with Short Bowel Syndrome. *J. Pediatr.* **2021**, *230*, 76–83.e75. [CrossRef] [PubMed]
13. Poindexter, B.B.; Martin, C.R. Impact of Nutrition on Bronchopulmonary Dysplasia. *Clin. Perinatol.* **2015**, *42*, 797–806. [CrossRef] [PubMed]
14. Gewolb, I.H.; Sobowale, B.T.; Vice, F.L.; Patwardhan, A.; Solomonia, N.; Reynolds, E.W. The Effect of Severe Intraventricular Hemorrhage on the Biorhythms of Feeding in Premature Infants. *Front. Pediatr.* **2021**, *9*, 673152. [CrossRef]
15. Qiu, W.; Chen, H.; Dincer, A.B.; Lundberg, S.; Kaeberlein, M.; Lee, S.I. Interpretable machine learning prediction of all-cause mortality. *Commun. Med.* **2022**, *2*, 125. [CrossRef] [PubMed]
16. Grinsztajn, L.; Oyallon, E.; Varoquaux, G. Why do tree-based models still outperform deep learning on typical tabular data? In Proceedings of the 36th Conference on Neural Information Processing Systems (NeurIPS 2022), New Orleans, LA, USA, 28 November–9 December 2022; New Orleans Convention Center: New Orleans, LA, USA, 2022.
17. Ke, G.; Meng, Q.; Finley, T.; Wang, T.; Chen, W.; Ma, W.; Ye, Q.; Liu, T.-Y. LightGBM: A Highly Efficient Gradient Boosting Decision Tree. In Proceedings of the 31st Conference on Neural Information Processing Systems (NIPS 2017), Long Beach, CA, USA, 4–9 December 2017.
18. Cawley, G.C.; Talbot, N.L.C. On Over-fitting in Model Selection and Subsequent Selection Bias in Performance Evaluation. *J. Mach. Learn. Res.* **2010**, *11*, 2079–2107.
19. Hastie, T.; Tibshirani, R.; Friedman, J.H. *The Elements of Statistical Learning: Data Mining, Inference, and Prediction*, 2nd ed.; Springer: New York, NY, USA, 2009; p. 745.
20. Molnar, C. Interpretable Machine Learning: A Guide for Making Black Box Models Explainable. 2019. Available online: https://leanpub.com/interpretable-machine-learning (accessed on 8 June 2024).

21. Lundberg, S.M.; Lee, S.-I. A Unified Approach to Interpreting Model Predictions. In Proceedings of the 31st Conference on Neural Information Processing Systems (NIPS 2017), Long Beach, CA, USA, 4–9 December 2017.
22. Brion, L.P.; Rosenfeld, C.R.; Heyne, R.; Steven Brown, L.; Lair, C.S.; Heyne, E.; Dohoney, E.L.; Burchfield, P.J.; Caraig, M. Association of age of initiation and type of complementary foods with body mass index and weight-for-length at 12 months of age in preterm infants. *J. Perinatol.* **2020**, *40*, 1394–1404. [CrossRef] [PubMed]
23. Spiegler, J.; Eisemann, N.; Ehlers, S.; Orlikowsky, T.; Kannt, O.; Herting, E.; Göpel, W. Length and weight of very low birth weight infants in Germany at 2 years of age: Does it matter at what age they start complementary food? *Eur. J. Clin. Nutr.* **2015**, *69*, 662–667. [CrossRef]
24. Morgan, J.B.; Lucas, A.; Fewtrell, M.S. Does weaning influence growth and health up to 18 months? *Arch. Dis. Child.* **2004**, *89*, 728–733. [CrossRef]
25. Baldassarre, M.E.; Di Mauro, A.; Pedico, A.; Rizzo, V.; Capozza, M.; Meneghin, F.; Lista, G.; Laforgia, N.; On behalf of Italian Society of Pediatrics (SIP), Italian Society of Neonatology (SIN), Italian Society of Pediatric Gastroenterology, Hepatology, and Nutrition (SIGENP) and Italian Federation of Paediatricians (FIMP). Weaning Time in Preterm Infants: An Audit of Italian Primary Care Paediatricians. *Nutrients* **2018**, *10*, 616. [CrossRef]
26. Ribas, S.A.; de Rodrigues, M.C.C.; Mocellin, M.C.; Marques, E.S.; da Rosa, G.P.C.; Maganha, C.R. Quality of complementary feeding and its effect on nutritional status in preterm infants: A cross-sectional study. *J. Hum. Nutr. Diet.* **2021**, *34*, 3–12. [CrossRef] [PubMed]

nutrients

MDPI

Article

Proactive Use of a Human Milk Fat Modular in the Neonatal Intensive Care Unit: A Standardized Feeding Protocol

Amanda Salley [1,*] and Martin L. Lee [2,3]

1 Hinsdale Hospital, UChicago Medicine AdventHealth Hinsdale Hospital, Hinsdale, IL 60521, USA
2 Prolacta Bioscience, City of Industry, CA 91746, USA
3 Department of Biostatistics, UCLA Fielding School of Public Health, Los Angeles, CA 90025, USA
* Correspondence: amandasalley.rd@gmail.com

Abstract: An exclusive human milk diet (EHMD) and standardized feeding protocols are two critical methods for safely feeding very low birth weight (VLBW) infants. Our institution initiated a standardized feeding protocol for all VLBW infants in 2018. In this protocol, a human milk fat modular was used only reactively when an infant had poor weight gain, fluid restriction, or hypoglycemia. As part of our NICU quality improvement program, internal utilization review data revealed a potential opportunity to improve growth and reduce costs. While maintaining the EHMD, a simple feeding guideline process change could provide cost savings without sacrificing caloric density or growth. We examined this process change in pre-post cohorts of VLBW infants. Methods: Our revised feeding protocol, established in October 2021, called for a human milk fat modular (Prolact CR) to be added to all infant feeding when parenteral nutrition (PN) and lipids were discontinued. The human milk fat modular concentration is 4 mL per 100 mL feed, providing approximately an additional 2 kcal/oz. We tracked data to compare (1) the use of the human milk fat modular, (2) the use of the human milk +8 fortifier, (3) overall growth before and after feeding protocol changes, and (4) cost differences between protocols. Results: Thirty-six VLBW infants were followed prospectively upon the introduction of the revised feeding protocol. In the revised era, the need for human milk +8 fortifier decreased from 43% to 14%. The decrease in the cost of a more costly fortifier provided a cost savings of USD 2967.78 on average per infant. Overall growth improved from birth to discharge, with severe malnutrition declining from 3.3% to 2.7% and moderate malnutrition declining from 37% to 8%. Conclusions: With the proactive use of a human milk fat modular in a standardized feeding protocol, our VLBW infants showed improved growth, lower malnutrition rates, and decreased use of higher caloric fortifiers.

Keywords: exclusive human milk diet; human milk fat modular; growth; standardized feeding protocol; NICU; cost savings

1. Introduction

An exclusive human milk diet (EHMD) utilizing a mother's own milk (MOM) with human milk-based fortifiers and the use of a standardized feeding protocol are two of the most evidence-based methods for decreasing necrotizing enterocolitis, improving growth, and decreasing overall co-morbidities in very low birth weight (VLBW) infants [1–3]. The American Academy of Pediatrics (AAP) and Surgeon General recommend the use of human milk for premature infants [3]. Huston et al. suggested that early fortification with an EHMD can improve growth and significantly decrease NEC (necrotizing enterocolitis).

The UChicago Medicine AdventHealth Hinsdale neonatal intensive care unit (NICU) initiated the use of an EHMD (Prolacta Bioscience, Inc., Duarte, CA, USA) in January 2016 for all VLBW infants. At the Level III NICU, an EHMD was provided to all infants born with £1500 g and/or £32 0/7 weeks at birth. In 2018, a standardized feeding protocol was implemented in an attempt to improve growth and maintain consistency within the

NICU. After an internal review of these NICU's VON (Vermont Oxford Network) data in 2019, it became apparent that NEC had risen beyond typical percentiles. Most infants affected were prenatally diagnosed with end diastolic flow, maternal abruption, metabolic acidosis at birth, and/or IUGR (intrauterine growth restriction). In 2020, an updated standard (Figure 1) and modified feeding protocol (Figure 2) were implemented to improve outcomes in our VLBWs.

Standard Feeding Protocol

Indications: All infants ≤1500 grams and/or 32 0/7 weeks

	Birthweight	Feeding Day 1	2	3	4	5	6	7	8	9	10	11
A	<500gm	1 mL q6h	1 mL q6h	1 mL q6h	1 mL	1.5 mL 2 mL	2.5 mL 3 mL	3.5 mL 4 mL	4.5 mL 5 mL	5.5 mL 6 mL	6.5 mL 7 mL	**
B	501-650 gm	1 mL	1 mL	1 mL	2 mL	2.5 mL 3 mL	3.5 mL 4 mL	4.5 mL 5 mL	5.5 mL 6 mL	6.5 mL 7 mL	7.5 mL 8 mL	**
C	651-850 gm	2 mL	2 mL	2 mL	3 mL	4 mL 5 ml	6 mL 7 mL	8 ml 9 mL	10 mL 11 mL	**		
D	851-1000 gm	2 mL	2 mL	2 mL	4 mL	5.5 ml 7 mL	8.5 mL 10 mL	11.5 mL 13 mL	**			
E	1001-1250 gm	3 mL	3 mL	5 mL	7 mL 9 mL	11 mL 13 mL	15 mL 17 mL	19 mL 21 mL	**			
F	1251-1500 gm	3 mL	5 mL	7 mL 9 mL	11 mL 13 mL	15 mL 17 mL	19 mL 21 mL	**Continue advancing BID to goal of 160 mL/kg				

Use Protocol F if baby is ≤32 weeks but over 1500 grams
Feeds given q3h unless otherwise noted
Fortify with Prolact +4
Fortify with Prolact +6

Figure 1. 2020 Standard Feeding Protocol.

Modified Feeding Protocol

Indications: Prenatally diagnosed end diastolic flow, IUGR and metabolic acidosis

	Birthweight	Feeding Day 1	2	3	4	5	6	7	8	9	10	11	12
A	<500gm	1 mL q6h	1 mL q6h	1 mL q6h	1 mL	1 mL	1.5 mL 2 mL	2.5 mL 3 mL	3.5 mL 4 mL	4.5 mL 5 mL	5.5 mL 6 mL	6.5 mL 7 mL	**
B	501-650 gm	1 mL	1 mL	1 mL	1.5 mL	2 mL	2.5 mL 3 mL	3.5 mL 4 mL	4.5 mL 5 mL	5.5 mL 6 mL	6.5 mL 7 mL	7.5 mL 8 mL	**
C	651-850 gm	2 mL	2 mL	2 mL	2 mL	3 mL	4 mL 5 mL	6 ml 7 mL	8 ml 9 mL	10 mL 11 mL	12 mL 13 mL	**	
D	851-1000 gm	2 mL	2 mL	2 mL	2 mL	3.5 mL	5 mL 6.5 mL	8 ml 9.5 mL	11 ml 12.5 mL	14 mL 15.5 mL	**		
E	1001-1250 gm	3 mL	3 mL	3 mL	5 mL	7 mL 9 mL	11 mL 13 mL	15 mL 17 mL	19 mL 21 mL	**			
F	1251-1500 gm	3 mL	3 mL	5 mL	7 mL 9 mL	11 mL 13 mL	15 mL 17 mL	19 mL 21 mL	**Continue advancing by 2 mL BID to goal of 160 mL/kg				

Use Protocol F if baby is ≤32 weeks but over 1500 grams
Feeds given q3h unless otherwise noted
Fortify with Prolacta +4
Fortify with Prolacta +6

Figure 2. 2020 Modified Feeding Protocol.

Within the updated 2020 feeding protocol, two significant changes were made. All infants born at £1000 g followed a protocol in which feeds were fortified with Prolact +4 (Prolacta Bioscience, City of Industry, CA, USA) or RTF 24 kcal/oz when feeds were at trophic level, less than 20–30 mL/kg. All infants following the feeding protocols received fortification of Prolact +6 (Prolacta Bioscience, City of Industry) or RTF 26 kcal/oz when feeds reached 60 mL/kg. Early fortification supports optimal growth and allows improved nutrients to be administered enterally rather than parenterally [1,4,5]. The second notable change coincides with parenteral nutrition (PN) guidelines. The updated feeding protocol identifies when enteral feeds are included in the total fluid volume. Utilizing the appropriate feeding protocol with early fortification and PN guidelines [1,5], both metabolic acidosis and metabolic bone disease can be avoided [6,7].

In October 2021, a chart review was conducted to determine the success of the current feeding protocol. Through this review, it was determined that 77% of infants utilized a human milk fat modular, and 43% of infants required a human milk +8 fortifier in addition to the fat modular. As part of our NICU quality improvement program, internal utilization review data revealed a potential opportunity to improve growth and reduce costs. Our hypothesis was that proactive use of a human milk fat modular, given after providing appropriate enteral protein intake, could improve growth and decrease cost through less use of the more costly +8 human milk fortifier.

Based on previous studies, it is known that fat loss is high in human milk feedings given via tube feeding [8,9]. The Rogers et al. study resulted in 6 ± 2% loss of fat via gravity feeds, 13 ± 3% loss of fat via pumps, and 40 ± 3% loss of fat in continuous feeds [8]. Once enteral protein needs are met (3.5–4.5 g/kg/day), proactively utilizing a human milk fat modular might allow for the provision of sufficient nutrients for optimal growth without the need for a more costly fortifier. Knake et al. found that 73% of infants in their study required the use of a cream supplement for weight gain <15 g/kg/d [9]. Tabata et al. displayed benefits from a human milk-derived fortifier and a human milk fat modular to improve infant weight gain with bioactive elements from mother's milk and increased fat delivery [10]. Hair et al. hypothesized that premature infants who receive an EHMD with a human milk fat modular would have weight gain at least as good as infants receiving a standard feeding regimen that consisted of MOM or donor HM with a human milk-derived fortifier [11]. Within this study, the authors found a significant enhancement in the growth of preterm infants who received a human milk fat modular in conjunction with an EHMD.

Proactive use of a human milk fat modular can meet the additional needs required by VLBW infants rather than waiting for reactive usage when growth remains poor. Utilizing an EHMD with reactive fat modular use, this Level III NICU was seeing moderate malnutrition indicators in 36.6% of infants (Table 1).

Table 1. Malnutrition severity (birthweight to discharge weight), following Goldberg et al. indicators.

Severity	Group 1 $n = 30$	Group 2 $n = 36$
No Malnutrition	14	16
Mild Malnutrition	4	16
Moderate Malnutrition	11	3
Severe Malnutrition	1	1

$p = 0.0061$ (exact chi-square test).

Following the chart review, the standardized feeding protocol adopted a more proactive use of the human milk fat modular (Figure 3).

Standard Feeding Protocol
Indications: All infants ≤1500 grams and/or 32 0/7 weeks

Birthweight		Feeding Day 1	2	3	4	5	6	7	8	9	10	11
A	<500gm	1 mL q6h	1 mL q6h	1 mL q6h	1 mL	1.5 mL 2 mL	2.5 mL 3 mL	3.5 mL 4 mL	4.5 mL 5 mL	5.5 mL 6 mL	6.5 mL 7 mL	**
B	501-650 gm	1 mL	1 mL	1 mL	2 mL	2.5 mL 3 mL	3.5 mL 4 mL	4.5 mL 5 mL	5.5 mL 6 mL	6.5 mL 7 mL	7.5 mL 8 mL	**
C	651-850 gm	2 mL	2 mL	2 mL	3 mL	4 mL 5 ml	6 mL 7 mL	8 ml 9 mL	10 mL 11 mL	**		
D	851-1000 gm	2 mL	2 mL	2 mL	4 mL	5.5 ml 7 mL	8.5 mL 10 mL	11.5 mL 13 mL	**			
E	1001-1250 gm	3 mL	3 mL	5 mL	7 mL 9 mL	11 mL 13 mL	15 mL 17 mL	19 mL 21 mL	**			
F	1251-1500 gm	3 mL	5 mL	7 mL 9 mL	11 mL 13 mL	15 mL 17 mL	19 mL 21 mL	**Continue advancing BID to goal of 160 mL/kg				

Use Protocol F if baby is ≤32 weeks but over 1500 grams
Feeds given q3h unless otherwise noted
Fortify with Prolact +4
Fortify with Prolact +6
Fortify with Prolact CR
Included in total fluids

Figure 3. 2021 Standard Feeding Protocol.

2. Materials and Methods

In implementing the new standardized feeding protocol (Figure 3), the human milk fat modular (Prolact CR, Prolacta Bioscience, City of Industry) is added to all feedings at 110–120 mL/kg when PN/lipids are discontinued. Prolact CR (Prolacta Bioscience, City of Industry) is mixed at 4 mL per 100 mL feeding, providing an average of an additional 2 kcal/oz. Enteral feeds are given over 30 min, unless a physician order is provided for an increased duration. Although not a preferred practice, when continuous feeds are needed in Level III, Prolact CR (Prolacta Bioscience, City of Industry) is provided in the same ratio, divided, and given as a bolus every 4 h before a new feeding syringe is placed [8].

Data were collected by the neonatal dietitian to compare the use of the human milk fat modular, the use of the human milk +8 fortifier, and overall growth.

2.1. *Inclusion/Exclusion Criteria*

This quality improvement study was approved by the institutional review board for UChicago Medicine AdventHealth Hinsdale, a level III NICU in the west suburbs of Chicago. The data were obtained through a retrospective review (births from January 2021 to October 2021) and ongoing data collection post-protocol changes (births from November 2021 to June 2022). Infants utilizing the EHMD—all infants born at £1500 and/or £32 weeks—were included in the study. Group 1 was obtained from a retrospective review, and Group 2 was established following the timing of protocol changes. Infants with presumed milk protein allergy, necrotizing enterocolitis (NEC) or intestinal perforation, transportation to an outside hospital, or need for an early wean were excluded from data collection (Table 2). This quality improvement initiative was to evaluate infants who could stay on a standardized feeding guideline and did not require a modified approach.

Table 2. Exclusions.

Exclusion Criteria	Group 1	Group 2
Presumed milk protein allergy	3	1
Nil per os (NPO)/Gastroschisis	1	--
Necrotizing enterocolitis	1	--

Table 2. *Cont.*

Exclusion Criteria	Group 1	Group 2
Spontaneous intestinal perforation	1	--
Early wean	--	1
Transport	1	1

2.2. Feeding Protocols

All infants were provided with an EHMD via a standardized feeding protocol (Figure 1). MOM was provided when available, or pasteurized donor human milk (DHM) was provided if the mother's own milk was unavailable. Infants' feeds were fortified with an exclusive human milk fortifier to 26 kcal/oz when feeds reached 60 mL/kg within the feeding protocol advancement. If an infant was experiencing poor weight gain, defined as <15 gm/kg/day, a human milk fat modular was added to feeds to provide an additional 2 kcal/oz. Furthermore, if weight velocity was still not meeting adequate levels (<15 g/kg/day), the human milk fortifier was further advanced to 28 kcal/oz, in addition to the human milk fat modular. Our goal growth velocity is 15–20 g/kg/day based on the most up-to-date literature for preterm infants <2 kg [12]. Group 1 was fed an EHMD with reactive use of a human milk fat modular when poor weight gain was seen over 3–4 days and further advanced to 28 kcal/oz if needed to support optimal weight gain [12].

In Group 2, all infants were provided an EHMD via the updated standardized feeding protocol (Figure 3) with the addition of prophylactic use of a human milk fat modular, Prolact CR (Prolacta Bioscience, City of Industry), when feeds were at 110–120 mL/kg. This addition of the human milk fat module coincided with the discontinuation of parenteral nutrition and lipids. The human milk fat modular was provided in the same ratio as Group 1. Infants were provided with 4 mL cream for every 100 mL of fortified feed. If an infant within Group 2 was seen to have poor weight gain (<15 g/kg/day), the human milk fortifier was advanced to 28 kcal/oz.

Group 1 and Group 2 were both weaned off an exclusive human milk diet when the infant was both 33 3/7 weeks and 1500 g. The wean occurred over a 4-day period and was complete when the infant was 34 0/7 weeks.

2.3. Data Collection

Data were collected by the neonatal dietitian from admission to discharge on all infants in Group 1 and Group 2. In addition to birth anthropometrics and gestational age, the use of human fat milk modular and 28 kcal/oz fortification were collected. Anthropometrics were assessed at 34 weeks, 36 weeks, and/or discharge. Z-scores were calculated using the 2012 Fenton growth curves via the online database PediTools. The database was utilized to plot all anthropometric measurements of individual patients [13]. This tool was used to report percentiles and z-scores with an integrated gestational age calculator. Secondary data collection was obtained through the Vermont Oxford Network database based on individual outcomes within Group 1 and Group 2. Supplemental data included length of stay, length of PN days, and malnutrition criteria.

2.4. Malnutrition Analysis

Per Goldberg et al., the primary indicators of neonatal malnutrition include a decline in weight for age z-score of 0.8–1.2 standard deviation (SD) for mild malnutrition, >1.2–2 SD in moderate malnutrition, and >2 SD in severe malnutrition [13].

2.5. Statistical Analysis

For categorical (qualitative) data, e.g., Vermont Oxford Index data or ethnicities, the comparison between the study groups used either the chi-square or Fisher's exact tests. For the latter, an exact p-value calculation was used for tables with small expected frequencies (<5). For quantitative data, e.g., length of stay, the Wilcoxon rank-sum test was used. All statistical comparisons were performed at a 5% significance level.

3. Results

There were no significant differences in infant demographics or ethnicities between the groups (Table 3). Group 1 had a mean gestational age of 29.0 weeks, and Group 2 had a mean gestational age of 29.5 weeks ($p = 0.34$). The mean birth weight in Group 1 was 1.24 kg, and in Group 2, it was 1.28 kg ($p = 0.6$). A variety of ethnicities were represented in both groups. Group 1 included 30 infants, 16 males and 14 females. Group 2 included 36 infants, 22 males and 14 females.

Table 3. Demographics/Ethnicities/Race.

Parameter	Group 1 $n = 30$	Group 2 $n = 36$	p-Value *
Sex (F)	14/30 (46.7%)	14/36 (38.9%)	0.52
Gestational age	29.0 ± 2.1 **	29.5 ± 1.9	0.34
Birthweight	1.24 ± 0.28	1.28 ± 0.28	0.60
Weight z-score birth	0.09 ± 0.76	−0.003 ± 0.86	0.82
Length z-score birth	−0.24 ± 0.93	−0.03 ± 1.05	0.27
Head circumference z-score birth	0.008 ± 0.99	−0.14 ± 0.89	0.64
Non-Hispanic White	13	13	0.55
Non-Hispanic Black or African American	5	8	0.57
Hispanic	9	7	0.32
Non-Hispanic Asian	0	4	0.12
Non-Hispanic, Other	3	4	1.0
Inborn	23	27	0.88

F = female; * sex, race/ethnicity, and location are analyzed by chi-square test; all others analyzed by Wilcoxon rank-sum test; ** mean ± SD.

Group 1 was fed an EHMD with the reactive use of a human milk fat modular. In Group 1, human milk +8 fortifier was used in 43.8% of the infants. Group 2 was fed an EHMD with the proactive use of a human milk fat modular. In Group 2, utilizing the updated feeding protocol with proactive human milk fat modular, the use of human milk +8 fortifier decreased to 13.9%. Group 1 required significantly more human milk +8 fortifier than Group 2 ($p = 0.0075$).

Both groups within the study utilized MOM as well as ready-to-feed donor milk-fortified products. There was no significant difference in the use of MOM between the groups ($p = 0.47$). In Group 1, 70% of infants had MOM fortified with Prolact +6. Of these infants, 43% required advancement to Prolact +8. Thirty percent (30%) of infants had no MOM and were fed with a Ready to Feed (RTF) 26 kcal/oz product. Of these infants, 44% required advancement to RTF 28 kcal/oz. All infants in Group 1 were fortified with the human milk fat modular prior to advancing calorie fortification. In Group 2, 78% of infants had MOM fortified with Prolact +6. Of these infants, 18% required advancement to Prolact +8. Twenty-two percent of infants had no MOM and were fed with RTF at 26 kcal/oz. Of these infants, none required advancement to RTF 28 kcal/oz. All infants in Group 2 were proactively fortified with human milk fat modular within the standard feeding protocol.

Utilizing neonatal malnutrition indicators [14], severe malnutrition declined from 3.3% to 2.7% from Group 1 to Group 2, respectively, while moderate malnutrition declined from 36.6% to 8.3% from Group 1 to Group 2. Overall, z-score weight change improved from birth to discharge within malnutrition severity ($p = 0.0061$) (Table 4).

The change in z-score birthweight to weight at 36 weeks improved for all infants in Group 2 (Table 3). Group 1's mean change in z-score was −0.91 ± 0.42, and Group 2's mean change in z-score was −0.80 ± 0.47 ($p = 0.4$). An overall improvement of 0.11 within the SD change was seen. The change in z-score birthweight to discharge weight improved by 0.17 SD ($p = 0.41$). The change in z-score birth length to discharge did not improve between Group 1 and Group 2. Of note, the median z-score length was much higher in Group 1 than in Group 2 ($p = 0.43$). The change in z-score head circumference to discharge improved by 0.21 SD ($p = 0.53$) in Group 2.

Table 4. Change in z-score.

Parameter	Group 1: Mean ± SD (Median)	Group 2: Mean ± SD (Median)	*p*-Value (Wilcoxon Rank-Sum Test)
Δ weight z: 36 weeks	−0.91 ± 0.63 (−0.88)	−0.76 ± 0.54 (−0.76)	0.21
Δ weight z: discharge	−0.94 ± 0.60 (−0.82)	−0.77 ± 0.57 (−0.81)	0.41
Δ length z: 36 weeks	−0.81 ± 0.93 (−0.70)	−0.82 ± 0.99 (−0.78)	0.83
Δ length z: discharge	−0.57 ± 1.01 (−0.63)	−0.71 ± 0.97 (−0.84)	0.43
Δ head circumference z: 36 weeks	−0.53 ± 0.85 (−0.56)	−0.48 ± 0.66 (−0.46)	0.82
Δ head circumference z: discharge	−0.44 ± 0.77 (−0.52)	−0.23 ± 0.98 (−0.38)	0.53

Weight change from birth to discharge improved within the two groups ($p = 0.0061$, Wilcoxon rank-sum test). In Group 1, 12 infants (40%) did not meet malnutrition criteria, with a difference in z-score for birthweight of 34 weeks. In Group 2, 17 infants (47.2%) did not meet malnutrition criteria, with a difference in z-score for birthweight of 34 weeks. The length change from birth to discharge was similar between the two groups ($p = 0.91$). In Group 1, 17 infants (56.6%) did not meet malnutrition criteria, with a difference in z-score for birth length to discharge. In Group 2, 18 infants (50%) did not meet malnutrition criteria with a difference in z-score for birth length to discharge. Head circumference, monitoring change in birth to discharge z-score, was stable within the two groups ($p = 0.31$). In Group 1, 19 infants (63.3%) did not meet malnutrition criteria with a difference in z-score for birth head circumference to discharge. In Group 2, 24 infants (66.6%) did not meet malnutrition criteria with a difference in z-score for birth head circumference to discharge. Of note, malnutrition was not an official ICD-10 code for billing purposes. Malnutrition indicators were utilized as a data collection standard.

3.1. Secondary Outcomes

Level III's average length of stay (LOS) before an EHMD was established in 2014–2015 was 77.2 days (Table 5). The use of an EHMD diet initially led to a decrease of 2.2 days (averaging 75 days) in 2016–2017. Developing and revising the standardized feeding protocol, in addition to the use of a proactive fat modular in the standardized NICU feeding protocol, resulted in an even further decrease in LOS. Group 1 average LOS was 66.2 days, with a further decrease of 2.4 days (averaging 63.8 days) in Group 2. The average number of days on PN was similar between the groups ($p = 0.73$). Group 1 had an average of 8.3 ± 3.5, and Group 2 had an average of 8.2 ± 4.0.

Participation in the VON provided an additional review of co-morbidities between Group 1 and Group 2. The incidence of chronic lung disease decreased between Group 1 and Group 2. Overall, the percentage of infants with CLD decreased from 40% in Group 1 to 19.4% in Group 2 ($p = 0.10$).

Table 5. Secondary outcomes.

Parameter	Group 1	Group 2	*p*-Value *
Length of stay	66.3 ± 25.7 ** median = 61.5	63.9 ± 26.2 median = 63	0.83
Total parenteral nutrition (days)	8.3 ± 3.5 median = 7.5	8.2 ± 4.0 median = 7.0	0.73

* analyzed by Wilcoxon rank-sum test; ** mean ± SD.

3.2. Cost Analysis

All infants in Group 1 and Group 2 were assessed daily based on products utilized and feeding volume provided over a 24 h period. Based on the cost of fortification in 2022, the total cost of fortification in Group 1 was, on average, USD 14,748.13 per infant (n = 30). The total cost of fortification in Group 2 was, on average, USD 11,780.35 per infant (n = 36). Most infants were on multiple products while following the EHMD. Table 6 displays the use of each product, including how many infants utilized the specific product. The overall amount of product used and volume received were divided by 100 mL to obtain the number of bottles needed (10 mL per Prolact CR used). The total price was then divided by the number of infants utilizing that product to provide the resulting cost per infant on average. Based on this information, the total cost of fortification in Group 1 was USD 442,443.86, with an average of USD 14,748.13 per infant. The total cost of fortification in Group 2 was USD 424,092.42, with an average of USD 11,780.35 per infant. A total savings of USD 18,351.44 was seen in Group 2 (Table 6). The average cost saved per infant was USD 2967.78.

Table 6. Cost Analysis.

Product Used	Group 1 (n = 30)	Group 2 (n = 36)	Cost Difference
Prolact +6 # of Infants	USD 73,856.77 21 * (USD *3516.99*)	USD 55,775.69 28 * (USD *1991.99*)	−USD 18,081.08
Prolact +6 w/Prolact CR # of infants	USD 95,833.69 18 * (USD *5324.09*)	USD 253,141.80 28 * (USD *9040.78*)	USD 157,308.11
Prolact +8 w/Prolact CR # of infants	USD 105,208.32 9 * (USD *11,689.81*)	USD 47,541.01 5 * (USD *9508.20*)	−USD 57,667.31
RTF 26 kcal/oz # of infants	USD 49,213.40 9 * (USD *5468.16*)	USD 4004.64 8 * (USD *500.58*)	−USD 45,208.76
RTF 26 kcal/oz w/Prolact CR # of infants	USD 34,867.07 6 * (USD *5811.18*)	USD 63,629.28 8 * (USD *7953.66*)	USD 28,762.21
RTF 28 kcal/oz w/Prolact CR # of infants	USD 83,464.61 4 * (USD *20,866.15*)	-- 0 (USD *0.00*)	−USD 83,464.61
Totals	** USD **442,443.86**	** USD **424,092.42**	** −USD **18,351.44**

* Cost of product used per infant. ** Total cost of product used per group.

Based on the Prolacta Bioscience 2022 Price List, the list price of Prolact +6 is USD 193.13. With the addition of Prolact CR at the ratio of 4 mL per 100 mL feed, the total cost is USD 209.61. The list price of Prolact +8 is USD 257.50. With the addition of Prolact CR, the total cost is USD 273.98. The cost savings of utilizing MOM with Prolact +6 with the addition of Prolact CR vs. Prolact +8 with the addition of Prolact CR (at 4 mL/100 mL) is USD 64.37 per 100 mL. The list price of ready-to-feed (RTF) 26 kcal/oz is USD 206.00. With the addition of Prolact CR at the ratio of 4 mL per 100 mL fee, the total cost is USD 222.48. The list price of RTF 28 kcal/oz is USD 267.80. With the addition of Prolact CR, the total cost is USD 284.28. If a ready-to-feed product is required, the cost savings is USD 61.80 per 100 mL. The list price for 2022 will differ based on the institution's purchasing price for fortification through contract negotiations.

Group 1 had an average length of stay of 66.2 days. The length of stay decreased to 63.8 days in Group 2, an average decrease of 2.4 days. Assuming a day of admission in the NICU costs an average of USD 3500 [15], the savings from the length of stay alone amounted to USD 8400 per infant.

4. Discussion

This quality improvement study displays significant cost savings with a reduction in the use of a +8 human milk fortifier by providing a human milk fat modular proactively within a standardized feeding protocol. An EHMD diet, the recommended nutrition source for all preterm infants, was utilized in both Group 1 and Group 2 within this study [16]. Recent studies have reported significant fat loss while utilizing MOM and donor human milk [8] feedings within this population. This study reviews the importance of a human milk fat module within a standard feeding protocol.

Target growth goals were maintained and improved in Group 2, as displayed by a change in z-score from birthweight to 36 weeks ($p = 0.21$) and birthweight to discharge ($p = 0.41$) (Table 3). The change in z-score from birth length to 36 weeks remained stable ($p = 0.83$). Group 2 also displayed improvements in head circumference z-score change from birth to 36 weeks ($p = 0.82$) and birth to discharge ($p = 0.53$).

Adhering to the standardized feeding protocol (Figure 3) assures all VLBW infants are meeting estimated nutrient needs, specifically for calories, protein, and fat, both enterally and parenterally. When a VLBW infant is receiving 160 mL/kg on average, the infant is being provided ~149 kcal/kg (varies due to the mother's own milk as well as fat loss) and 4.32–4.48 gm/kg protein. Based on Koletzko's latest recommendations [17], enteral protein is being met within these guidelines. Additional calories from fat are needed to support optimal growth. This quality improvement study showed improved growth and cost savings when a human milk fat modular was used earlier within a standardized feeding protocol.

Belfort and Ehrenkranz suggested that both greater weight gain, while supporting linear growth and head growth, and the use of human milk fortifiers can be associated with better neurodevelopmental outcomes [18]. Early fortification and proactive use of cream are two essential components of a successful feeding protocol utilizing an EHMD [1].

Limitations to this study include the use of a retrospective cohort study design with uncontrolled changes within the NICU during the two periods of data collection. The small sample size likely contributed to the lack of statistical significance. Excluded infants were removed from data collection if the feeding protocol was not being followed (Table 4).

5. Conclusions

The proactive use of a human milk fat modular within a standardized protocol for all VLBW infants demonstrated a significant reduction in the use of a high-calorie human milk fortifier, which in turn provided cost savings to the NICU. Proactive use of a human milk fat modular supported appropriate growth, trending toward improvement. Further cost savings were potentially found because of decreased co-morbidities and decreased length of stay. Further study is needed to confirm these findings in other institutions due to the small study size.

Author Contributions: Conceptualization, A.S.; methodology, A.S.; software, A.S.; validation, A.S. and M.L.L.; formal analysis, M.L.L.; investigation, A.S.; resources, A.S.; data curation, A.S.; writing—original draft preparation, A.S.; writing—review and editing, A.S. and M.L.L.; visualization, A.S.; supervision, A.S.; project administration, A.S. All authors have read and agreed to the published version of the manuscript.

Funding: This research received no external funding.

Institutional Review Board Statement: The study was conducted in accordance with the Declaration of Helsinki, and approved by the Institutional Review Board of AdventHealth (approval code 1901083-1, 13 April 2022).

Informed Consent Statement: Patient consent was waived due to retrospective review and ongoing data collection post protocol changes.

Data Availability Statement: The original contributions presented in the study are included in the article, further inquiries can be directed to the cooresponding author.

Conflicts of Interest: Amanda Salley is on the Speaker's Bureau for Prolacta Bioscience but was not a consultant for the company when the data for this study were being collected. Martin L. Lee is an employee of Prolacta Bioscience but was brought on to the manuscript after the study data were collected.

References

1. Huston, R.K.; Markell, A.M.; McCulley, E.A.; Gardiner, S.K.; Sweeney, S.L. Improving Growth for Infants ≤1250 Grams Receiving an Exclusive Human Milk Diet. *Nutr. Clin. Pract.* **2018**, *33*, 671–678. [CrossRef]
2. Kaplan, H.; Poindexter, B. Standardized Feeding Protocols: Evidence and Implementation. In *Nutritional Care of Preterm Infants: Scientific Basis and Practical Guidelines*; Karger AG: Basel, Switzerland, 2021; pp. 289–300.
3. Hair, A.B.; Peluso, A.M.; Hawthorne, K.M.; Perez, J.; Smith, D.P.; Khan, J.Y.; O'Donnell, A.; Powers, R.J.; Lee, M.L.; Abrams, S.A. Beyond Necrotizing Enterocolitis Prevention: Improving Outcomes with an Exclusive Human Milk-Based Diet. *Breastfeed. Med.* **2016**, *11*, 70–74, Correction in *Breastfeed. Med.* **2017**, *12*, 663. [CrossRef]
4. Cormack, B.E.; Jiang, Y.; Harding, J.E.; Crowther, C.A.; Bloomfield, F.H.; ProVIDe Trial Group. Neonatal Refeeding Syndrome and Clinical Outcome in Extremely Low-Birth-Weight Babies: Secondary Cohort Analysis from the ProVIDe Trial. *J. Parenter. Enter. Nutr.* **2021**, *45*, 65–78. [CrossRef]
5. Robinson, D.T.; Calkins, K.L.; Chen, Y.; Cober, M.P.; Falciglia, G.H.; Church, D.D.; Mey, J.; McKeever, L.; Sentongo, T. Guidelines for parenteral nutrition in preterm infants; the American Society for Parenteral and Enteral Nutrition. *J. Parenter. Enter. Nutr.* **2023**, *47*, 830–858. [CrossRef] [PubMed]
6. Bonsante, F.; Gouyon, J.B.; Robillard, P.Y.; Gouyon, B.; Iacobelli, S. Early optimal parenteral nutrition and metabolic acidosis in very preterm infants. *PLoS ONE* **2017**, *12*, e0186936. [CrossRef] [PubMed]
7. Faienza, M.F.; D'Amato, E.; Natale, M.P.; Grano, M.; Chiarito, M.; Brunetti, G.; D'Amato, G. Metabolic Bone Disease of Prematurity: Diagnosis and Management. *Front. Pediatr.* **2019**, *7*, 143. [CrossRef] [PubMed]
8. Rogers, S.P.; Hicks, P.D.; Hamzo, M.; Veit, L.E.; Abrams, S.A. Continuous feedings of fortified human milk lead to nutrient losses of fat, calcium and phosphorous. *Nutrients* **2010**, *2*, 230–240. [CrossRef] [PubMed]
9. Knake, L.A.; King, B.C.; Gollins, L.A.; Hurst, N.M.; Hagan, J.; Ford, S.L.; Hair, A.B. Optimizing the Use of Human Milk Cream Supplement in Very Preterm Infants: Growth and Cost Outcomes. *Nutr. Clin. Pract.* **2020**, *35*, 689–696. [CrossRef] [PubMed]
10. Tabata, M.; Abdelrahman, K.; Hair, A.B.; Hawthorne, K.M.; Chen, Z.; Abrams, S.A. Fortifier and cream improve fat delivery in continuous enteral infant feeding of breast milk. *Nutrients* **2015**, *7*, 1174–1183. [CrossRef] [PubMed]
11. Hair, A.B.; Bergner, E.M.; Lee, M.L.; Moreira, A.G.; Hawthorne, K.M.; Rechtman, D.J.; Abrams, S.A.; Blanco, C.L. Premature Infants 750–1250 g Birth Weight Supplemented with a Novel Human Milk-Derived Cream Are Discharged Sooner. *Breastfeed. Med.* **2016**, *11*, 133–137. [CrossRef] [PubMed]
12. Section of Neonatology, Department of Pediatrics, Baylor College of Medicine. *Guidelines for Acute Care of the Neonate*, 29th ed.; 2021–2022; Section of Neonataolgy, Department of Pediatrics, Baylor College of Medicine: Houston, TX, USA, 2021.
13. Chou, J.H.; Roumiantsev, S.; Singh, R. PediTools Electronic Growth Chart Calculators: Applications in Clinical Care, Research, and Quality Improvement. *J. Med. Internet Res.* **2020**, *22*, e16204. [CrossRef] [PubMed]
14. Goldberg, D.L.; Becker, P.J.; Brigham, K.; Carlson, S.; Fleck, L.; Gollins, L.; Sandrock, M.; Fullmer, M.; Van Poots, H.A. Identifying Malnutrition in Preterm and Neonatal Populations: Recommended Indicators. *J. Acad. Nutr. Diet.* **2018**, *118*, 1571–1582. [CrossRef] [PubMed]
15. Black, L.; Hulsey, T.; Lee, K.; Parks, D.C.; Ebeling, M.D. Incremental hospital costs associated with comorbidities of prematurity. *Manag. Care* **2015**, *24*, 54–60. [CrossRef] [PubMed]
16. American Acadmey of Pediatrics Section on Breastfeeding. Breastfeeding and the use of human milk. *Pediatrics* **2012**, *129*, e827–e841. [CrossRef] [PubMed]
17. Koletzko, B.; Fheah, F.C.; Domellof, M.; Poindexter, B.; Vain, N.; Goudoever, J. (Eds.) Nutritional care of preterm Infants: Scientifiic basis and practical guidelines. In *World Review of Nutrition and Dietetics*, 2nd ed.; Karger AG: Basel, Switzerland, 2021.
18. Belfort, M.B.; Ehrenkranz, R.A. Neurodevelopmental outcomes and nutritional strategies in very low birth weight infants. *Semin. Fetal Neonatal Med.* **2017**, *22*, 42–48. [CrossRef] [PubMed]

nutrients

Article

Bioelectrical Impedance in Premature Newborns and Its Relationship with Diet Therapy in a Neonatal Intensive Care Unit

Catiuscie Cabreira da Silva Tortorella [1,*], Bárbara Mendes Paz Chao [2], Estela Iraci Rabito [3], Mônica Nunes Lima [1] and Ana Lúcia Figueiredo Sarquis [1]

[1] Graduate Program in Child and Adolescent Health, Department of Pediatrics, Hospital de Clínicas, Federal University of Paraná, Curitiba 80060-900, Brazil; moica.lima.ufpr@gmail.com (M.N.L.); ana.sarquis@hc.ufpr.br (A.L.F.S.)
[2] Department of Pharmacy, Midwest State University, Guarapuava 80060-900, Brazil; bmpaz@unicentro.br
[3] Department of Nutrition, Federal University of Paraná, Curitiba 80060-900, Brazil; estelarabito@yahoo.com.br
* Correspondence: cathicabreira@hotmail.com; Tel.: +55-(42)-991425354

Abstract: (1) Background: To estimate resistance, reactance, and phase angle values among moderate preterm infants and their variation according to neonatal and maternal characteristics and nutritional intake. (2) Methods: This was a cohort that evaluated 43 moderate preterm infants using bioelectrical impedance analysis. The study variables included resistance, reactance, and phase angle measurements, in addition to classification of nutritional intake. (3) Results: Mean resistance was $602.0 \pm 118.2\ \Omega$, reactance was $57.2\ \Omega$ (IQR = 42.6–65.2), and phase angle was 522° (IQR = 4.1–6.6). Lower resistance values were found in the presence of risky pregnancy ($532.2 \pm 111.9\ \Omega$ vs. $650.9 \pm 97.9\ \Omega$, $p < 0.001$) and lower reactance values, in the presence of harmful maternal lifestyle habits at both the first ($p = 0.01$) and second assessments ($p = 0.01$). Eight preterm infants were considered to have insufficient nutritional intake (23.5%); 17, sufficient (50.0%) and 9, partially sufficient (26.5%). There was less reactance among preterm infants with insufficient nutritional intake ($p < 0.001$). (4) Conclusions: The bioelectrical impedance analysis measurements were within the range of values reported in other studies. There was an association between full diet and adequate nutritional intake with higher resistance values, while a lower reactance value was associated with the presence of risky pregnancy and harmful maternal lifestyle.

Keywords: electrical impedance; nutrition therapy; preterm birth; maternal behaviors; intensive care units; neonatal

1. Introduction

Providing adequate nutrition to promote the growth and development of premature newborns is one of the major challenges facing health workers in Neonatal Intensive Care Units (NICUs) on a daily basis [1]. Food intolerance is typical in this gestational age and becomes even more severe in more premature infants, and it often prevents full enteral nutrition [2]. Some of the clinical signs of food intolerance are the presence of more than 50% of gastric residuals of the diet previously offered, regurgitation, abdominal distension, and/or emesis. Such signs indicate the inability of the immature gastrointestinal tract to receive enteral nutrition. The consequences are well known and have a great impact on the survival of these babies, namely the prolonged use of alternative feeding routes, including total parenteral nutrition and its related risks, as well as longer hospital stays and nutritional failure [3]. Although this is a common problem, the growth of preterm infants has been monitored through anthropometric measurements since the 18th century [4]. Additionally, body composition measurements are not part of their routine care, although it is known that they can optimize nutrition and help promote better neonatal growth and development [1].

Premature babies initially present growth restriction, and later they show catch-up growth with higher fat mass (FM) accumulation, although they present lower fat-free mass gain (FFMG), higher adiposity, and lower linear growth in the first two years of life compared to full-term infants. Adequate nutrition in the NICU has been associated with weight gain and fat-free mass (FFM). These two indices are associated with better cognition in childhood, and FFM is a better predictor of neurodevelopment than weight [1].

The problem lies in defining the best method of measuring body composition in babies, for the sake of greater precision, fewer adverse effects, and feasibility. Many of them, such as magnetic resonance and nuclear spectroscopy, dual-energy X-ray absorption (DEXA), and isotopic techniques, have already been studied and have limitations, either in terms of cost, impossibility of being performed at the bedside, or exposure to ionizing radiation [1]. The most promising methods are air displacement plethysmography (ADP), bioelectrical impedance (BIA), skinfold caliper measurements, and ultrasonography. Only ADP is valid for use in term infants, but it has a high cost and cannot be used in severely ill babies. The performance of ultrasound measurements may be affected by tissue compression and may lead to inconsistent results, while skinfold measurements are influenced by hydroelectrolytic status and have low accuracy. BIA is a noninvasive, radiation-free method that can be performed at the bedside, which shows promising results for the assessment of body composition of newborn (NBs) [1]. BIA measurements help determine the values of resistence (R), reactance (Xc), and phase angle (PA), and the values of R are inversely proportional to the quantity of intra- and extracellular fluids [5]. Muscle tissues present lower R, owing to the good electric current conduction favored by a large amount of water and electrolytes. On the other hand, the adipose tissue is not a good conductor of electric current because of the low amount of water and electrolytes; thus, R is higher. Based on these indicators, it can be inferred that higher values of R indicate a greater amount of adipose tissue and a lower amount of muscle tissue [5]. The Xc values indicate the presence of a healthy or disease-affected membrane; it can be used as a prognostic marker in different clinical situations, and low values may be associated with poor nutritional status [6,7]. However, further studies are needed to better understand the values and behavior of BIA in this age group. Therefore, the objective of the study was to use BIA to collect data on R, Xc, and AF values in the first days of life, taking into account neonatal and maternal factors.

2. Materials and Methods

This was a cohort conducted in an NICU of a tertiary university hospital in southern Brazil, between April 2018 and December 2021. The following inclusion criteria were considered: preterm infants (gestational age < 37 weeks) in need of intensive care and whose mothers agreed to participate in the study by signing an informed consent form. The following exclusion criteria were adopted: (a) preterm infants with congenital malformation and genetic syndromes; (b) cases of technical problems and errors in BIA measurements; (c) preterm infants in a warm cradle at the time of measurement; (d) unavailable limb areas for electrode placement, either by the presence of a catheter or by any other impediment; (e) preterm infants transferred to another hospital or their homes; (f) absence of complete follow-up; or (g) absence of complete nutritional information.

Of 129 preterm infants initially eligible for the study, 79 were excluded because of transfer (3), inaccurate or wrong BIA measurements (7), moist-cradle bed (4), unavailable area for placement of BIA electrodes (4), absence of complete follow-up (44), and absence of complete nutritional information (17). There were a total of 50 eligible babies: 43 moderate preterm infants (≥32 and <34 gestational weeks), 4 extremely preterm infants (<28 gestational weeks), and 3 late preterm NBs (≥34 and <37 gestational weeks), who were excluded from the study owing to the small number of cases. Thus, there were 43 moderate preterm infants left, whose BIA measurement was performed in the first and second weeks of life (Figure 1).

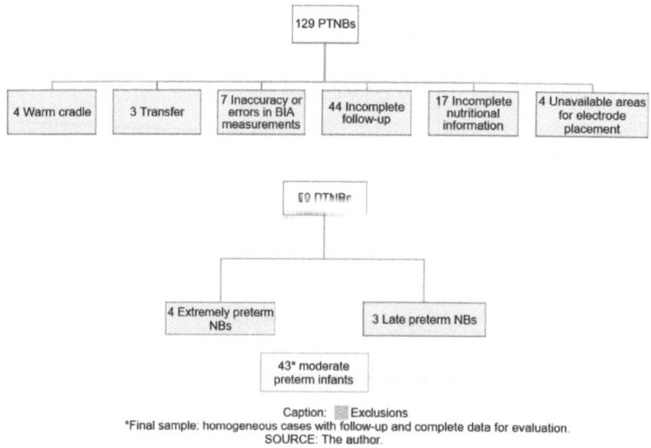

Figure 1. Flowchart for sample selection. PTNBs: preterm infants; BIA: bioelectrical impedance; NBs: newborn.

The study variables included BIA measurements, particularly R and Xc (in ohms (Ω)), nutritional intake, classified as sufficient (energy ≥ 100 kcal/kg/day and protein > 3 g/kg/day), partial (energy ≥ 100 kcal/kg/day and protein ≤ 3 g/kg/day or <100 kcal/kg/day and protein > 3 g/kg/day) or insufficient (energy < 100 kcal/kg/day and protein ≤ 3 g/kg/day) [8]. It is worth noting that, at the first moment of the evaluation, 79.1% (34) of all evaluated NBs were fed human milk or pasteurized colostrum from the Human Milk Bank; 16.3% (7) were on parenteral nutrition and 4.7% (2) were fed nutrient formula for high-risk newborns for human milk addition. In the second evaluation, 46.5% (20) were fed pasteurized human milk, 12 pasteurized human milk with FM85®, 27.9% (5) pasteurized human milk supplemented with infant formula for premature babies, and another 27.9% (5) were fed only infant formula.

R, Xc, and PA values were determined with a Bio Scan Maltron 916 (50 kHz) bioelectrical impedance analyzer, Maltron internacional, Reino Unido, Rayleigh [9]. The test was always carried out on the same day of the week, before the newborn was fed the next diet, and in two moments: between the 1st and 7th days of life and between the 8th and 15th days of life. The test was performed with preterm infants in the supine position, in the incubator itself, and with electrodes positioned on the same side of the body (right or left, chosen according to the area available), with a spacing of at least four centimeters between them, measured with a sterile body tape measure. The electrodes were cut in half lengthwise to fit the placement area on the limb of the preterm infants, and they were not reused [10]. All procedures were performed after the materials had been sanitized with 70% isopropyl alcohol.

The weight of the newborns was measured on a Filizola® (Filizola Baby, São Paulo, Brazil) calibrated pediatric scale, with a minimum capacity of 125 g and a maximum capacity of 15 kg. The preterm infants were evaluated while being naked and positioned on the scale so that their body weight was distributed over the surface. For measuring length, standardization was followed in the NICU using the Frankfurt plan [11]. Information on dietary therapy, neonatal data (sex, gestational age, birth weight, nutritional status, 1 min and 5 min Apgar scores. And respiratory distress syndrome), and maternal characteristics (age, schooling, smoking/alcoholism/drug addiction, number and type of deliveries, neonatal appointments, previous diseases, diseases during pregnancy, twin pregnancy, and risky pregnancy) were collected from the medical records of the preterm infants. Women were considered to have risk pregnancies when aged 35 years or older and/or with comorbidities such as excess weight, systemic arterial hypertension, diabetes mellitus, and thyroid disease. Dietary evolution was recorded for each NB from the beginning of the diet until they reached the energy target of 100/kcal/kg/day and of 3 g/kg/day proteins [8].

The minimum sample size was estimated considering the type II error of 10%, significance level of 5%, and effect magnitude of 60 points on average, indicating a minimum sample of 40 cases, with a test power of 95%.

To estimate the difference between continuous variables with symmetric distribution, the t-test was applied for dependent samples, followed by analysis of variance (one-way ANOVA), while asymmetric distribution was checked by the Wilcoxon test and Kruskal–Wallis ANOVA, followed by Duncan and Mann–Whitney post hoc tests, respectively. For all tests, a significance level of 5% was considered (Statistics 4.0 (StatSoft Power Solutions, Inc., Palo Alto, CA, USA). The study was approved by the Research Ethics Committee of the institution under registration number 4.640.434.

3. Results

The study sample consisted of 43 moderate preterm infants, born to adult women with a mean age of 30.5 ± 7.7 years, 20 of whom (58.8%) were considered to have risk pregnancies (Table 1).

Table 1. Maternal characteristics.

Characteristics	Mean ± SD/n (%)
Age (years)	30.5 ± 7.7
Education—High School	25 (73.5%)
Smoking/Alcoholism/Drug addiction	7 (20.6%)
Primiparous	4 (11.8%)
Abortions	14 (41.2%)
Number of prenatal appointments <6	2 (2.9%)
Previous diseases	18 (52.9%)
Diseases during pregnancy	15 (44.1%)
Cesarean section	18 (52.9%)
Twin pregnancy	14 (41.2%)
Risky pregnancy	20 (58.8%)

SD = Standard deviation/risky pregnancy = Maternal age > 35 years and/or disease during pregnancy.

The mean gestational age of the preterm infants was 33.3 ± 0.6 weeks, and birth weight was 1997.9 ± 542.0 g, with a proportional distribution of cases in relation to sex (1:1.2) (Table 2).

Table 2. Characteristics of premature newborns.

Characteristics	Mean ± SD/n (%)
Sex (M/F)	25/21 (53.5%/46.5%)
Gestational age (weeks)	33.3 ± 0.6
Birth weight (grams)	1997.9 ± 542.0
Nutritional status	
SGA	4 (11.8%)
SUGA	26 (76.4%)
LGA	4 (11.8%)
1 min Apgar score	
<3	4 (11.8%)
4–6	4 (11.8%)
<6	26 (76.4%)
5 min Apgar score	
<6	34 (100.0%)
RDS	24 (70.6%)

SUGA = suitable for gestational age, F = female, LGA = large for gestational age, M = male, SGA = small for gestational age, RDS = respiratory distress syndrome.

In the first assessment, carried out in a median of 2 days (interquartile range/IQR = 2–2), all preterm infants had not been fed a full diet, and their nutritional intake was lower than desirable for growth (<100 Kcal/kg/day and <3 g/kg/day of protein). Table 3 shows the BIA measurements of preterm infants in the first assessment.

Table 3. Resistance, reactance, and phase angle measurements in the first assessment of bioelectrical impedance.

Measures	Mean SD/Median (IQR)
R (Ω)	602.0 ± 118.2
Xc (Ω)	57.2 (42.6–65.2)
PA (°)	5.22 (4.1–6.6)

PA = phase angle, R = resistance, Xc = reactance, IQR = interquartile variation.

Regarding maternal characteristics, there was a lower R value in the presence of risky pregnancy (532.2 ± 111.9 Ω vs. 650.9 ± 97.9 Ω, $p < 0.001$). In addition, there were lower Xc values in the preterm infants of pregnant women with harmful life habits (smoking, alcoholism, drug addiction) in both the first (42.6 ± 14.2 Ω vs. 61.3 ± 17.6 Ω, $p = 0.01$) and in the second assessments (44.0 ± 9.2 Ω vs. 65.1 ± 20.1 Ω, $p = 0.01$).

For neonatal characteristics, there was no significant difference between the measurements of BIA according to gender, nutritional status, 1 min Apgar score, or Respiratory Distresss Syndrome (RDS) ($p > 0.05$ for all study variables).

The second assessment was carried out in a median of 8 days (interquartile range/IQR = 8–9), and 8 preterm infants were considered to have insufficient nutritional intake (23.5%), while 17 had sufficient (50.0%), and 9 had partially sufficient (26.5%) intakes, respectively.

There was a significantly lower R among preterm infants with insufficient intake when compared to those with sufficient or partially sufficient intake ($p < 0.001$), and there was no significant difference for Xc and PA (Table 4).

Table 4. Resistance, reactance, and phase angle measurements according to nutritional intake in the second week of life.

Measures	Insufficient NI (n = 8)	Partial NI (n = 9)	Sufficient NI (n = 17)	*p*
R (Ω)	540.0 ± 120.2	611.2 ± 105.1	731.5 ± 101.9	<0.001 [1]
Xc (Ω)	53.6 (375–83.5)	54.4 (49.2–86.9)	54.9 (49.5–74.5)	0.47 [2]
PA (°)	5.3 (3.6–9.3)	5.4 (4.1–7.2)	4.3 (3.8–5.9)	0.53 [2]

PA = phase angle, NI = nutritional intake, R = resistance, Xc = reactance. [1] One-way ANOVA, Duncan post hoc test, [2] Kruskal–Wallis ANOVA.

There was also a significant increase in R between the first and second weeks of assessment only among preterm infants with sufficient nutritional intake (Table 5).

Table 5. Resistance, reactance, and phase angle measurements in the first and second assessments according to nutritional intake.

	Measures	1st Assessment	2nd Assessment	*p*
	R (Ω)	48.0 ± 114.7	540.0 ± 120.2	0.15 [1]
Insufficient NI (n = 8)	Xc (Ω)	56.5 (42.3–69.6)	55.1 (42.9–87.5)	0.61 [2]
	PA (°)	5.9 (5.0–7.8)	5.3 (3.9–11.5)	0.73 [2]
	R (Ω)	618.4 ± 146.0	61.2 ± 105.1	0.86 [1]
Partial NI (n = 9)	Xc (Ω)	65.1 (61.7–8.6)	56.2 (50.0–63.2)	0.46 [2]
	PA (°)	6.6 (4.4–6.9)	4.5 (4.1–5.0)	0.68 [2]
	R (Ω)	637.1 ± 69.0	731.5 ± 102.0	<0.001 [1]
Sufficient NI (n = 17)	Xc (Ω)	49.0 (41.7–59.7)	54.9 (49.5–70.2)	0.14 [2]
	PA (°)	4.5 (3.7–5.2)	4.3 (4.0–5.4)	0.77 [2]

PA = phase angle, NI = nutritional intake, R = resistance, Xc = reactance, [1] Dependent t-test, [2] Wilcoxon test.

4. Discussion

In the present study, the mean R in moderate preterm infants in the first days of life was 602.0 ± 118.2 Ω, Xc, with a median of 57.2 Ω (IQR = 42.6–65.2) and median PA of 5.22° (IQR = 4.1–6.6). There was a lower Xc value in the presence of risky pregnancy and harmful life habits and a higher R value among preterm infants on a full diet and with sufficient nutritional intake; these values increased between the 1st and 2nd assessments in this group of NBs. Such values are similar to those reported by Coradine et al. [12], who also studied moderate preterm infants, but they were lower than those found by Margutti et al. [13] with late preterm infants ($p < 0.001$) for R values. There was a significant variation in Xc values (48.7 to 67.9 Ω, $p < 0.001$), and PA values were similar (4.7 to 5.2°, $p > 0.05$). Piccoli et al. [10], Savino et al. [14], Margutti et al. [15], and Coradine et al. [12] reported a significant variation in the mean values of R (466.0 to 684.8 Ω, $p < 0.001$), Xc (22.0 to 50.3 Ω; $p < 0.001$), and PA (2.5 to 5.1°; $p < 0.001$), even though all those studies had focused on full-term newborns (Table A1). It is known that measurements of BIA may vary according to a number of factors, including age range, and among NBs, according to gestational age, clinical stability, fluid status, limb movement at the time of examination, and imminence of feeding time [1], which may account for differences in measures. Thus, there should be further studies on BIA, with a standardized design for newborns, to establish more accurate reference values in neonatology. Margutti et al. [15] also found that R values were significantly lower in male NBs, who seem to have a higher amount of FFM, total body water, and cell membranes. Nehab et al. [16] suggested that their postnatal growth is higher, with greater muscle mass gain. In the sample of the present study, there was no significant difference in the measurements of BIA between male and female NBs.

For maternal characteristics, there was a lower Xc value in the presence of a risky pregnancy in this study. Nehab et al. [16] reported that gestational factors, susceptible to prevention, influence the amount of mass of neonatal fat, while demographic characteristics (mother's age, gestational age, and NB's sex) affect the amount of FFM at birth. Using ADP, they found that maternal morbidities such as diabetes mellitus and systemic arterial hypertension during pregnancy determined a higher percentage of body fat in full-term NBs. FFM was also influenced by the newborn's sex, birth weight, gestational age, and maternal age.

It is known that the use of tobacco, drugs, and alcohol during pregnancy results in high rates of abortion, limited fetal growth, premature membrane rupture, and premature delivery. Other consequences include higher mortality rates and lower birth weight [17]. In the present study, a lower Xc value was found in preterm infants whose mothers smoked and/or used drugs and/or drank alcoholic beverages. Whereas Xc is an indicator of cellular membrane integrity and intra- and extracellular water distribution, this finding may be indicative of cellular death and/or decreased cellular integrity, and Xc can be used as a marker to determine the intensity of the harmful effects of tobacco, alcohol, and drugs on the body of NBs whose mothers had these habits during pregnancy. Zhou et al. [18] also reported the impact of smoking on the growth and body composition of NBs, possibly owing to higher energy expenditure, maternal malnutrition, lower maternal weight gain during pregnancy, placental dysfunction, and possible direct effects of tobacco on maternal and fetal metabolism.

Studies have shown that early deficits in protein and energy intake during the first two weeks of life affect neonatal growth and long-term neurocognitive development in infants [19–21]. Gerritsen et al. [22] found that only 58% of moderate preterm infants had the recommended protein intake on the seventh day of life and that the average increase of 1 g/kg/day in protein intake in the first week of life resulted in a significant increase in weight. Baillat et al. [23] also found that early energy and protein intake positively influences neonatal growth and that 60% of children did not have such nutritional intake at the end of the first week of life. They indicated that for every increase of 10 Kcal/kg/d at the end of the first week of life, delayed extrauterine growth was 27% less likely to occur in preterm infants (or = 0.73; 95% CI = 0.66–0.82). In the present sample, similarly, there was an increase in R only among preterm infants who were fed a full diet and had the recommended nutritional intake of proteins in the second week of life.

A randomized study was carried out in Spain with 38 non-breastfed preterm infants, who were divided into three groups and received different amounts of protein through infant formulas for preterm infants. The authors found that the groups that had received the highest amount of protein (4.2 g/kg/day or 4.7 g/kg/day) presented higher FFM gain than the control group with non-supplemented formula [23]. Mól et al. [24] assessed the difference in body composition between preterm infants fed breastmilk or infant formula in comparison to those born at term. They found that those fed formula presented higher R, which may represent a greater amount of adipose tissue and lower FFM. However, preterm infants fed breastmilk did not present differences in body composition in comparison with the control group of term infants.

Preterm infants have a high metabolic rate and biochemical immaturity, which affects metabolic functions and leads to high nutritional risk [25]. The assessment of nutritional status in the first days of life allows a better understanding of intrauterine development and enables better dietary therapy intervention [26], for the purpose of promoting neurodevelopment and reducing metabolic risks in the long term. Anthropometric assessments are used to measure weight loss in the first days of life and check if the subsequent weight gain and length are within the values considered normal according to the literature. However, this form of assessment does not distinguish between adipose, muscle, and physiological tissues [1,27].

BIA has been applied in research with preterm infants; however, there are issues to consider when using it in this age group. By means of a portable and easy-to-use device, BIA measures the opposition (or impedance) of the body tissues to the flow of an alternating, low-intensity electric current, which passes through the body through electrodes that are in contact with the skin. BIA measurements help determine the values of R, Xc, and PA [5]. Because PA indicates the presence of a healthy or disease-affected membrane, it can be used as a prognostic marker in different clinical situations, and low values may be associated with poor nutritional status [6,7]. In our study, despite the PA reduction behavior, this result was not significant, which contrasts with the findings of the studies by Margutti et al. [13] and Coradine et al. [12], who found a significant reduction. It is noteworthy that this behavior does not occur among full-term newborns, who present an increase in phase angle as the days go by [10,15]. This is possibly due to the difference in hydration status between premature and full-term newborns. Preterm infants have a higher body water content, and their body losses are more intense, reaching 15–20% of water in the first days of life. They also go through an intense extracorporeal adaptation period, with different types of evolution of fed diets. These factors may account for the reduction of the phase angle in premature infants. The clinical condition of newborns must also be considered, as the presence of diseases can interfere with cellular integrity, mass, and hydration, i.e., the prognostic values of the phase angle can also differ in groups of patients with different clinical conditions [6].

In the first two weeks of life of preterm infants, especially between the 4th and 9th days of life, there is a physiological loss of neonatal weight of up to 15% of body weight. After that period, there is a growth peak at a speed that attempts to reproduce intrauterine rates. Dietary therapy and assessment of body composition at these critical moments of oscillating weight deserve special attention in the quest for growth and nutritional quality [28].

Body weight in NICUs is the most common measure for nutritional assessment of preterm infants, although it does not evaluate body composition. With the increased survival of these NBs, there has been greater interest in nutritional assessment, as feeding in the first weeks of life has a direct impact on their development [29,30].

A limitation of the present study was that it did not estimate the nutritional intake of macronutrients. It is known that there is variation in breastmilk donors and interindividual variation and changes in the composition of breastmilk according to lactation stage, in addition to differences in macronutrients in milk and nutritional formulas used for supplementation. Therefore, it is impossible to assess the profile of each macronutrient.

It is worth pointing out that there was no attempt to insert the data found in prediction equations for FFM and FM, because the use of BIA in NBs needs to be standardized, and the existing equations have methodological limitations and still lack validation for Brazilian NBs [5].

5. Conclusions

The BIA measurements made in this sample are within the range of values reported in other studies with preterm infants and full-term NBs. There is considerable variation, which possibly reflects the lack of standardization in the design of studies using this method of assessing body composition in NBs. There was an association between full diet and adequate nutritional intake with higher R values, as well as a lower Xc value associated with the presence of a risky pregnancy and harmful life habits, such as smoking.

Author Contributions: Conceptualization, C.C.d.S.T. and A.L.F.S.; methodology, C.C.d.S.T., A.L.F.S. and E.I.R.; formal analysis, M.N.L.; investigation, C.C.d.S.T.; writing—original draft preparation, C.C.d.S.T.; writing—review and editing, C.C.d.S.T. and B.M.P.C.; supervision, A.L.F.S. and E.I.R.; All authors have read and agreed to the published version of the manuscript.

Funding: This research received no external funding.

Institutional Review Board Statement: The study was conducted in accordance with the Declaration of Helsinki, and approved by the Research Ethics Committee Hospital de Clínicas, Federal University of Paraná under registration number 4.640.434, 9 April 2021.

Informed Consent Statement: Informed consent was obtained from all subjects involved in the study.

Data Availability Statement: The data presented in this study are available upon request from the corresponding author (cathicabreira@hotmail.com or cathicabreira@gmail.com). The data is not publicly available, as it is the property of the Hospital de Clinica da UFPR.

Conflicts of Interest: The authors declare no conflict of interest.

Appendix A

Table A1. Resistance, reactance, and phase angle values in the literature.

Authors	Piccoli et al. [10]	Savino et al. [14]	Margutti et al. [15]	Margutti et al. [13]	Coradine et al. [12]	Coradine et al. [12]	Study Date
Town/City (Country)	Turin (Italy)	Turin (Italy)	Ribeirão Preto (Brazil)	Ribeirão Preto (Brazil)	Curitiba (Brazil)	Curitiba (Brazil)	Curitiba (Brazil)
n	163	58	109	68	76	17	43
GA	Term	Term	Term	35.0 ± 1.6	32.8 ± 2.6	37.9 ± 1.1	33.0 ± 0.6
R	505 ± 60	466 ± 64	684.8 ± 53.5	794.7 ± 124.3	569.0 ± 113.1	524.8 ± 96.2	602.0 ± 118.2
Xc	43 ± 14	22 ± 12	37.5 ± 5.3	67.9 ± 31.9	48.7 (26.6–103.2)	50.3 (18.5–93.4)	57.2 (42.6–65.2)
PA	4.86 ± 1	2.5 ± 1.5	3.14 ± 0.43	4.92 ± 2.18	4.7 (2.5–12.1)	5.1 (2.7–10.7)	5.2 (4.1–6.6)

References

1. Nagel, E.; Hickey, M.; Teigen, L.; Kuchnia, A.; Curran, K.; Soumekh, L.; Earthman, R.D.C.; Demerath, E.; Ramel, M.D.S. Clinical application of body composition methods in premature infants. *J. Parenter. Enteral. Nutr.* **2020**, *44*, 785–795. [CrossRef]
2. Evidence-Based Medicine Group, Neonatologist Society, Chinese Medical Doctor Association. Clinical guidelines for the diagnosis and treatment of feeding intolerance in preterm infants. *Zhonguo Dang Dai Er Ke Za Zhi* **2020**, *22*, 1047–1055. [CrossRef]
3. Yuan, Z.; Junmei, Y.; Wen, H.; Deng, X.; Li, X.; Su, S. Feeding intolerance alters the gut microbiota of preterm infants. *PLoS ONE* **2019**, *22*, e0210609. [CrossRef] [PubMed]
4. Ferreira, A.A. Evaluation of the Growth of Children: Path of the Growth Charts. *Demetra* **2012**, *7*, 191–202. Available online: https://www.e-publicacoes.uerj.br/index.php/demetra/article/viewFile/3786/4291 (accessed on 2 December 2022).

5. Kyle, U.G.; Bosaeus, I.; De Lorenzo, A.D.; Deurenberg, P.; Elia, M.; Gómes, J.M.; Heitmann, B.L.; Kent-Smith, L.; Melchior, J.C.; Pirlich, M.; et al. Bioelectrical impedance analysis—Part I: Review of principles and methods. *Clin. Nutr.* **2004**, *23*, 1226–1243. [CrossRef] [PubMed]
6. Bosy-Westhhal, A.; Danielzik, S.; Dörhöfer, R.-P.; Later, W.; Wiese, S.; Müller, M.J. Phase angle from bioelectrical impedance analysis: Population reference values by age, sex, and body mass index. *J. Parenter. Enter. Nutr.* **2006**, *30*, 309–316. [CrossRef] [PubMed]
7. Thibault, R.; Makhlouf, A.M.; Mulliez, A.; Gonzalez, M.C.; Kekstas, G.; Kozjek, R.N.; Preiser, J.C.; Isabel Ceniceros Rozalen, I.C.; Sylvain Dadet, S.; Krznaric, Z.; et al. Fat-free mass at admission predicts 28-day mortality in intensive care unit patients: The international prospective observational study Phase Angle Project. *Intensive Care Med.* **2016**, *42*, 1445–1453. [CrossRef] [PubMed]
8. Zigler, E.E.; Thureen, P.J.; Carlson, S.J. Agressive nutrition of the very low birthweight infant. *Clin. Perinatol.* **2002**, *29*, 225–244. [CrossRef]
9. Lyons-Reid, J.; Derraik, J.G.B.; Ward, L.C.; Tint, M.T.; Kenealy, T.; Cutfield, W.S. Bioelectrical impedance analysis for assessment of body composition in infants and young children-A systematic literature review. *Clin. Obes.* **2021**, *11*, e12441. [CrossRef]
10. Piccoli, A.; Fanos, V.; Peruzzi, L.; Schena, S.; Pizzini, C.; Borgione, S.; Bertino, M.D.E.; Chiaffoni, M.D.G.; Coppo, M.D.R.; Tatò, M.D.L. Reference values of the bioelectrical impedance vector in neonates in the first week after birth. *Nutrition* **2002**, *18*, 383–387. [CrossRef]
11. Brasil Ministério da Saúde. *Orientações para a Coleta e Análise de Dados Antropométricos em Serviços de Saúde: Norma Técnica do Sistema de Vigilância Alimentar e Nutricional—SISVAN*; Secretaria de Atenção à Saúde, Departamento de Atenção Básica: Brasília, Brazil, 2011; pp. 1–76. Available online: https://bvsms.saude.gov.br/bvs/publicacoes/orientacoes_coleta_analise_dados_antropometricos.pdf (accessed on 9 March 2019).
12. Coradine, A.V.P.; Lima, M.N.; Sarquis, A.L. Can phase angle in newborns at neonatal intensive care units be na indicator of mortality and prognosis? *J. Parenter. Enteral. Nutr.* **2021**, *45*, 1192–1196. [CrossRef]
13. Margutti, A.V.B.; Bustamante, C.R.; Sanches, M.; Padilha, M.; Beraldo, R.A.; Monteiro, J.P.; Camelo Júnior, J.S. Bioelectrical impedance vector analysis (BIVA) in stable preterm newborns. *J. Pediatr.* **2012**, *88*, 253–258. [CrossRef]
14. Savino, F.; Grasso, G.; Cresi, F.; Oggero, R.; Silvestro, L. Bioelectrical impedance vector distribution in the first year of life. *Nutrition* **2003**, *19*, 492–496. [CrossRef]
15. Margutti, A.V.B.; Monteiro, J.P.; Camelo, J.S. Reference distribution of the bioelectrical impedance vector in healthy term newborns. *Br. J. Nutr.* **2010**, *104*, 1508–1513. [CrossRef]
16. Nehab, S.R.G.; Villela, L.D.; Abranches, D.A.; Rocha, D.M.; Silva, L.M.D.L.; Amaral, Y.N.V.; Junior, S.C.G.; Soares, F.V.M.; Méio, M.D.B.B.; Moreira, M.D.L. Influence of gestational and neonatal factors on body composition of full-term newborns. *J. Pediatr.* **2020**, *96*, 771–777. [CrossRef]
17. Costa-Orvay, J.A.; Figueras-Aloy, J.; Romera, G.; Closa-Monasterolo, R.; Carbonell-Estrany, X. The Effects of Varying Protein and Energy Intakes on the Growth and Body Composition of Very Low Birth Weight Infants. *Nutr. J.* **2011**, *29*, 140. [CrossRef]
18. Zhou, S.J.; Hawke, K.; Collins, C.T.; Gibson, R.A.; Makrides, M. Does maternal smoking in pregnancy explain the diferences in the body composition trajectory between breastfeed and formula-fed infants? *Br. J. Nutr.* **2020**, *123*, 402–409. [CrossRef]
19. Martin, C.R.; Brown, Y.F.; Ehrenkranz, R.A.; O'Shea, T.M.; Allred, E.N.; Belfort, M.B.; McCormick, M.C.; Leviton, A. Extremely Low Gestational Age Newborns Study Investigators. Nutritional practices and growth velocity in the first month of life in extremely premature infants. *Pediatrics* **2009**, *124*, 649–657. [CrossRef]
20. Moyses, H.E.; Johnson, M.J.; Leaf, A.A.; Cornelius, V. Early parenteral nutrition and growth outcomes in preterm infants: A systematic review and meta-analysis. *Am. J. Clin. Nutr.* **2013**, *97*, 816–826. [CrossRef]
21. Embleton, N.E.; Pang, N.; Cooke, R.J. Postnatal malnutrition and growth retardation: An inevitable consequence of current recommendations in preterm infants? *Pediatrics* **2001**, *107*, 270–273. [CrossRef] [PubMed]
22. Gerritsen, L.; Lindeboom, R.; Hummel, T. Prescribed protein intake does not meet recommended intake in moderate- and late-preterm infants: Contribution to weight gain and head growth. *Nutr. Clin. Pract.* **2020**, *35*, 729–737. [CrossRef]
23. Baillat, M.; Pauly, V.; Dagau, G.; Berbis, J.; Boubred, F.; Fayol, L. Association of First-Week Nutrient Intake and Extrauterine Growth Restriction in Moderately Preterm Infants: A Regional Population-Based Study. *Nutrients* **2021**, *13*, 227. [CrossRef]
24. Mól, N.; Zasada, M.; Kwinta, P. Does type of feeding affect body composition in very low birth weight infants?—A prospective cohort study. *Pediatr. Neonatol.* **2019**, *60*, 135–140. [CrossRef]
25. Pereira, G.R.; Alves Filho, N.; Leone, C.R.; Trindade Filho, O. *Avaliação Nutricional no Recém-Nascido Pré-Termo*; Nutrição do Recém-Nascido Pré-Termo; MEDBOOK-Ed Científica Ltda: Rio de Janeiro, Brazil, 2008; pp. 241–261.
26. Pinto, E.; Oliveira, A.R.; Alencastre, H.; Lopes, C. Avaliação da Composição Corporal na Criança por Métodos não Invasivos. *Arq. Med.* **2005**, *19*, 47–54.
27. Roche, S.; Lara-Pompa, N.E.; Macdonald, S.; Fawbert, K.; Valente, J.; Williams, J.E.; Hill, S.; Wells, J.; Fewtrell, M. Análise vetorial de bioimpedância elétrica (BIVA) em crianças hospitalizadas; preditores e associações com desfechos clínicos. *Eur. J. Clin. Nutr.* **2019**, *73*, 1431–1440. [CrossRef]
28. Silveira, R.C.; Procianoy, R.S. Crescimento nos primeiros anos de vida de recém-nascidos de muito baixo peso. In *Programa de Atualização em Neonatologia (PRORN)—Sociedade Brasileira de Pediatria*; Procianoy, R.S., Leone, C.R., Eds.; Artmed/Panamericana: Porto Alegre, Brazil, 2003; pp. 49–86.

29. Tang, W.; Ridout, D.; Modi, N. Influence of respiratory distress syndrome on body composition after preterm birth. *Arch. Dis. Child.-Fetal Neonatal Ed.* **1997**, *77*, F28–F31. [CrossRef] [PubMed]
30. Rugolo, L.M.; Bentlin, M.R.; Rugolo Junior, A.; Dalben, I.; Trindade, C.E. Crescimento de prematuros de extremo baixo peso nos primeiros dois anos de vida. *Rev. Paul. Pediatr.* **2007**, *25*, 142–149. [CrossRef]

nutrients

MDPI

Article

Biochemical Profiling of Urine Metabolome in Premature Infants Based on LC−MS Considering Maternal Influence

Jeong-Hun Mok [1,†], Junhwan Song [2,†], Won-Ho Hahn [3], Seonghyeon Cho [4], Jong-Moon Park [4], Jiwon Koh [2], Ho Kim [2] and Nam Mi Kang [5,*]

1 Department of Medical Device Management and Research, Samsung Advanced Institute for Health Sciences and Technology (SAIHST), Sungkyunkwan University, Seoul 06355, Republic of Korea; jeonghunmok@naver.com

2 Department of Pediatrics, Soonchunhyang University Cheonan Hospital, Cheonan 31151, Republic of Korea; joonanii@schmc.ac.kr (J.S.); 99018@schmc.ac.kr (J.K.); c78141@schmc.ac.kr (H.K.)

3 Department of Pediatrics, Soonchunhyang University Seoul Hospital, Seoul 04401, Republic of Korea; dr.hahn.md@gmail.com

4 Basil Biotech, Incheon 22002, Republic of Korea; shcho@basilbiotech.com (S.C.); basil@basilbiotech.com (J.-M.P.)

5 Department of Nursing, Research Institute for Biomedical & Health Science, Konkuk University, Chungju-si 27478, Republic of Korea

* Correspondence: nmkang03@kku.ac.kr; Tel.: +82-43-450-3033; Fax: +82-43-840-3958

† These authors contributed equally to this work.

Abstract: In this study, Liquid Chromatography–Mass Spectrometry (LC-MS)-based metabolomics profiling was conducted to elucidate the urinary profiles of premature infants during early and late postnatal stages. As a result, we discovered significant excretion of maternal drugs in early−stage infants and identified crucial metabolites like hormones and amino acids. These findings shed light on the maternal impact on neonatal metabolism and underscore the beneficial effects of breastfeeding on the metabolism of essential amino acids in infants. This research not only enhances our understanding of maternal–infant nutritional interactions and their long−term implications for preterm infants but also offers critical insights into the biochemical characteristics and physiological mechanisms of preterm infants, laying a groundwork for future clinical studies focused on neonatal development and health.

Keywords: metabolomics; premature infant; maternal nutrition; LC−MS; human milk; neurotransmitter; amino acid

1. Introduction

The nutrition of premature infants is important for health, optimal growth, and development. The maternal nutritional status, as the primary source of nutrition, significantly influences the nutritional status of premature infants. Poor maternal nutrition can lead to negative birth outcomes, such as low birth weight, and has long−term postnatal effects [1]. Consequently, previous research has focused on maternal–infant substance transfer, including aspects of breastfeeding, obesity, and environmental factors [2–4]. Representatively, breastfeeding, as a primary method of nutrition, has been extensively studied for its impact on maternal diet, infant microbiota, and disease or allergy prevention [5,6]. The ongoing focus on mother–infant interaction research underscores its significance, especially for preterm infants with underdeveloped organs. However, previous studies on related biomolecules have often focused on key nutrients such as ATP, glucose, long−chain fatty acids, and amino acids [7–11]. This has led to a relative deficiency in the comprehensive analysis of postnatal metabolic mechanisms in premature infants. Therefore, our study uses metabolomics profiling of the urine of premature infants to elucidate the influence of maternal health on infant metabolic activity.

A previous study on this concept has identified early postnatal metabolic adaptation and maturation alterations, focusing largely on essential energy metabolic cycles using nuclear magnetic resonance (NMR) [12]. In contrast, our study applies Liquid Chromatography–Mass Spectrometry (LC−MS) analysis and provides novel insights into complex molecular mechanisms in premature infants, including a broader examination of breastfeeding and formula feeding. In particular, LC−MS, known for its higher sensitivity and wider metabolite detection range, has enhanced our in−depth exploration [13]. Many studies have collected samples within two weeks after birth as a baseline for investigating the effects of prematurity, and the first three days after birth are considered an extremely preterm period for the organs of infants including the brain [14–17]. We analyzed urine samples from the immediate postnatal period (1–3 days after birth) as the Early group and two weeks later (13–16 days after birth) as the Late group. Furthermore, we regrouped the samples into a human milk (HM) group and a formula milk feeding (FM) group to assess maternal influence and metabolic activity.

Through this approach, we have identified significant metabolic mechanisms in the urine of premature infants. This suggests new insights into how maternally transferred metabolites can influence neonatal status. Our non−invasive study contributes to a multifaceted evaluation of factors affecting premature infants, leading to a deeper understanding of their health and developmental processes.

2. Materials and Methods

2.1. Materials

High−performance liquid chromatography (HPLC)−grade water, acetonitrile (ACN), and methanol (MeOH) were purchased from JT Baker (Philipsburg, NJ, USA). Formic acid (FA) was purchased from Sigma−Aldrich (St. Louis, MO, USA).

2.2. Methods

2.2.1. Sample Preparation

Urine samples of 10 mL were collected from premature infants aged 1–3 days (Early group, n = 22) and 13–16 days (Late group, n = 12, serving as the control) and were subsequently stored at −20 °C. We also recorded whether the infants were fed breast milk (n = 14) or formula milk (n = 20). The HM group included infants who received 100% human milk and a mixed diet of breast and formula milk, with at least two−thirds breast milk, while the formula milk group consisted exclusively of infants fed 100% formula milk. The urine samples were thawed on ice and mixed with a four times larger volume of chilled MeOH. These mixtures were vortexed for 1 min, centrifuged gently, and then incubated overnight. After incubation, mixtures were centrifuged at 14,000× g for 10 min. The supernatants were then transferred to new tubes and dried. Finally, the samples were resuspended in 0.1% FA and prepared for LC−MS analysis.

2.2.2. LC−MS

Chromatographic separation of the samples was performed using an Agilent ZORBAX Eclipse Plus C18 Rapid Resolution High−Definition column (2.1 × 50 mm, 1.8 µm particles) on a Vanquish UHPLC system (Thermo Fisher Scientific, Waltham, MA, USA), equipped with a Q−Exactive Hybrid Quadrupole−Orbitrap MS (Thermo Fisher Scientific, Waltham, MA, USA). The mobile phases consisted of 0.1% FA in water (solvent A) and 0.1% FA in 80% ACN (solvent B), with a flow rate of 200 µL/min. The total gradient time was set at 30 min: 2.5% B for 0–2 min; 2.5–12% B for 2–11 min; 12–28% B for 11–15 min; 28–60% B for 15–22 min; 60–96% B for 22–22.1 min; 96% B for 22.1–24 min; 96–2.5% B for 24–24.1 min; and finally, 2.5% B for 24.1–30 min. Mass spectrometry was conducted in positive electrospray ionization mode, equipped with a Heated Electrospray Ionization Probe, with the resolutions for full−MS and MS/MS scans set at 70,000 and 17,500 (at 400 m/z), respectively. The scanning range was 100–1000 m/z, with an automatic gain control (AGC) target of 1×10^6, a maximum IT of 100 ms, and a normalized collision energy

(NCE) for dd−MS2 of 30%. For data analysis, Compound Discoverer 3.3.2.31 (Thermo Fisher Scientific, Waltham, MA, USA) was used; this workflow for untargeted metabolomics facilitated retention time alignment and compound identification. MzCloud was employed to annotate compounds at the MS/MS level. The ChemSpider, Human Metabolome Database (HMDB), and Kyoto Encyclopaedia of Genes and Genomes (KEGG) databases were utilized to annotate features based on exact mass, using the internal database of Compound Discoverer. Chemical background noise was eliminated using a blank file.

3. Results

We analyzed the metabolism of premature infants in early postnatal development to identify significant molecular mechanisms and enhance our understanding of the biological relationship between mothers and infants. In this study, a total of 34 urine samples were collected from premature infants, divided into two groups: an Early group consisting of infants within 1 to 3 days postnatal and a Late group consisting of infants within 13 to 16 days postnatal. We analyzed the data using LC−MS and applied the Metabolomics Society's Metabolomics Standards Initiative annotation for standardization [18]. The filtration was performed at Level 2, involving exact mass matching (10 ppm) and a fragmentation score over 80 in the mzCloud database. As a result, 316 metabolites were identified, and 284 metabolites were used in the final analysis, applying a data filtering process, which included using an interquartile range variance filter to exclude the least informative 10% of variables (Table S1).

3.1. Multivariate Analysis

Initially, a multivariate data matrix was simplified using Principal Component Analysis (PCA) to visualize similarities and differences between the two groups (Figure 1a). The PCA revealed that PC1 and PC2 accounted for 36.3% and 18.2% of the variance, respectively, distinguishing the groups overall, but some overlap in patterns was observed in certain samples. Subsequently, Partial Least Squares Discriminant Analysis (PLS-DA) indicated distinct pattern separation between the Early and Late groups, with the first two components explaining 35.4% and 10.4% of the variance, respectively (Figure 1b). In cross−validation, the model with five components achieved an accuracy of 1.0, an R^2 of 0.995, and a Q^2 of 0.907, confirming its high efficacy in differentiating the groups (Figure 1c). Notably, small−scale clustering patterns in each group were consistently observed in both the PCA and PLS−DA results. To identify the metabolites driving these patterns, the top 30 substances with a Variable Importance in Projection (VIP) score above 1 were selected (Figure 1d, Table S2). Remarkably, ampicillin was distinguished by a significantly higher VIP score compared to other metabolites, emerging as a key differentiator between the groups. Moreover, other drugs such as penicillin−V and pirbuterol, as well as common urinary metabolites like α−lactose and creatine, were also identified in Figure 1d. Then, metabolites associated with purine and pyrimidine metabolism, such as hypoxanthine, guanine, thiamine, and adenosine, were identified as significant. The multivariate analysis patterns were consistently reflected in a heatmap displaying the quantitative values of each compound (Figure 2a). The clear separation between the two groups and the small−scale clusters observed in the PCA and PLS−DA results were further clarified by visualizing the results in the heatmap. Specifically, in the Early group, certain metabolites showed quantitative values that indicated overlapping patterns between the groups, reconfirming the presence of overlaps and specific small−scale clusters. Therefore, additional analysis was performed to identify specific metabolites.

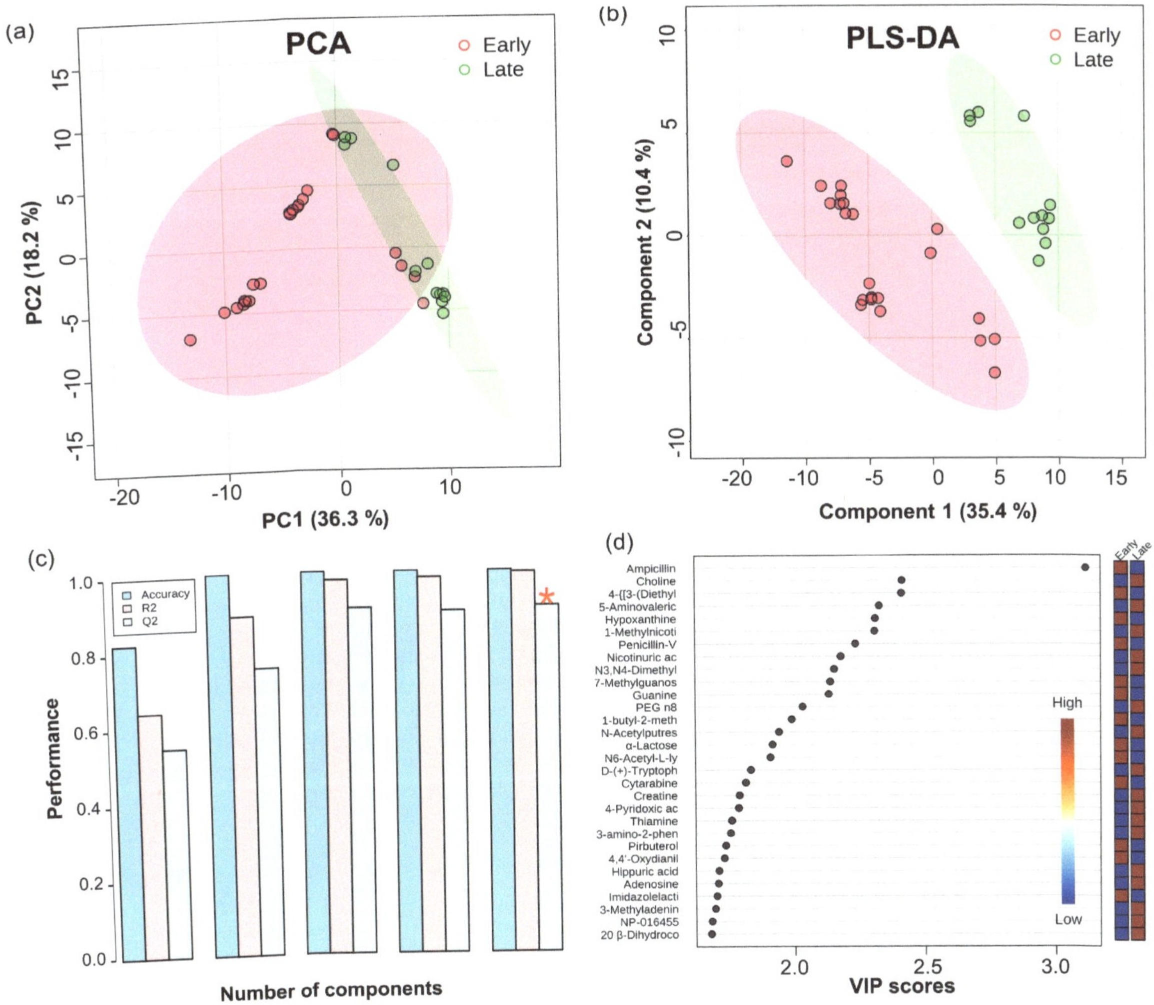

Figure 1. (**a**) PCA plot; (**b**) PLS−DA plot for Early group (red; $n = 22$) and Late group (green; $n = 12$); (**c**) Cross−validation of PLS−DA; (**d**) The top 30 VIP scores. The marker of (*) in (**c**) represents the highest value in the performance measure.

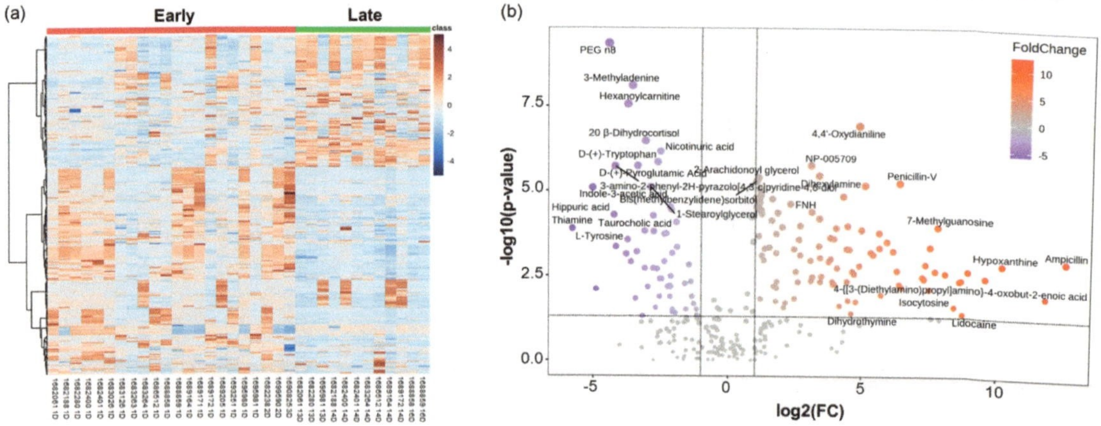

Figure 2. (**a**) Heatmap showing the quantitative values of each compound; (**b**) Volcano plot displaying DEMs (fold change (FC) = Early (n = 22)/late (n = 12), p−value < 0.05, $|\log2(FC)| > 1$).

3.2. *Differential Analysis*

After observing the differences in patterns between groups, differentially expressed metabolites (DEMs) were identified to specify the significant molecular mechanisms. As a result, 100 upregulated and 56 downregulated metabolites were identified in the Early group and visualized in a volcano plot (Figure 2b). Representatively, drug metabolites including ampicillin, penicillin−V, amoxicillin, and lidocaine were upregulated in the Early group. Furthermore, purine and pyrimidine metabolism-related metabolites (hypoxanthine, guanine, thiamine), steroid hormone−related metabolites (pregnenolone and 5α−pregnan−3,20−dione), and neurotransmitter metabolism−related metabolites (taurine, S−adenosylmethionine, and L−pyroglutamic acid) were mainly upregulated in the Early group. All DEMs are listed in Table S3.

3.3. *Univariate and Multivariate ROC Curve Analysis*

In parallel with other results used to select features in the data, we considered the quantitative regulation of metabolites, VIP scores, and receiver operating characteristic (ROC) curve analysis results. First, we elaborated and validated the significant DEMs through univariate ROC curve analysis. In order to select variables with high reliability, we applied the least absolute shrinkage and selection operator (LASSO) using the R package "glmnet" to select variables for model establishment (Table S4). This predictive model was computed using 10−fold cross−validation. As a result, we identified predictive models of nine metabolites with a good area under the ROC curve and CI, including penicillin−V (AUC = 0.96, CI = 0.86–1), steroid hormones such as 20β−dihydrocortisol (AUC = 0.96, CI = 0.83–1), and 17 α−hydroxyprogesterone (AUC = 0.91, CI = 0.78–0.99) (Figure 3, Table S5). Considering the complexity of the profiling results, driven by interactions among multiple metabolites as variables, an additional predictive model was constructed using multivariate ROC curve analysis to supplement a deeper understanding (Table S6). Multivariate ROC curve analysis was conducted based on Monte Carlo Cross−Validation. The classification method and the feature-ranking mechanism were performed using the PLS−DA algorithm. The ROC curves were generated for models with different numbers of features (5, 10, 15, 25, 50, 100), displaying plots, AUC values, and Cis. The AUC scores ranged from 0.905 to 0.986 (Figure 4a). In predictive accuracy, the 50−feature panel of model 5 achieved the highest accuracy, as shown in Figure 4b. However, to avoid overfitting, we selected the 10−feature panel of model 2. This decision was based on the AUC values exceeding 0.9 for all ROC curves and similar accuracy predictions for more than 10 features. The AUC of model 2 was 0.94 with a 95% CI of 0.687–1, visualized in

Figure 4c. From this predictive model, the most significant 10 markers were classified based on average importance (Figure 4d). Consistently, ampicillin, choline, and penicillin−V were significantly reaffirmed and validated in this model as significant metabolites, coordinating with other analytical results.

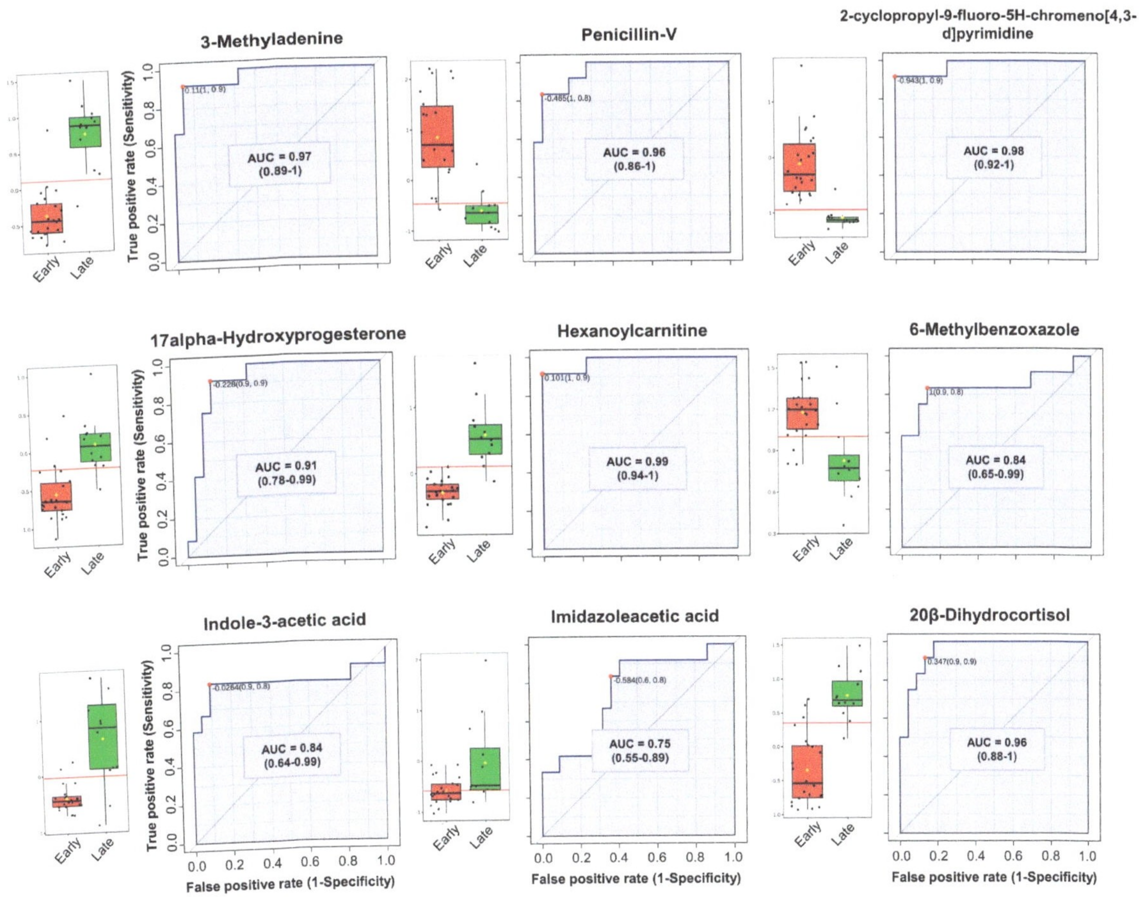

Figure 3. ROC curves with area under the ROC curve (AUC) and confidence interval (CI) values on selected DEMs using the least absolute shrinkage and selection operator (LASSO) feature selection algorithm. Boxplots of relative concentrations for selected DEMs between Early (red) and Late (green) groups. The black dots represent the concentrations of the selected feature from all samples. Horizontal red lines on the boxplot indicate the optimal cutoff. Yellow diamonds represent the mean concentration of each group.

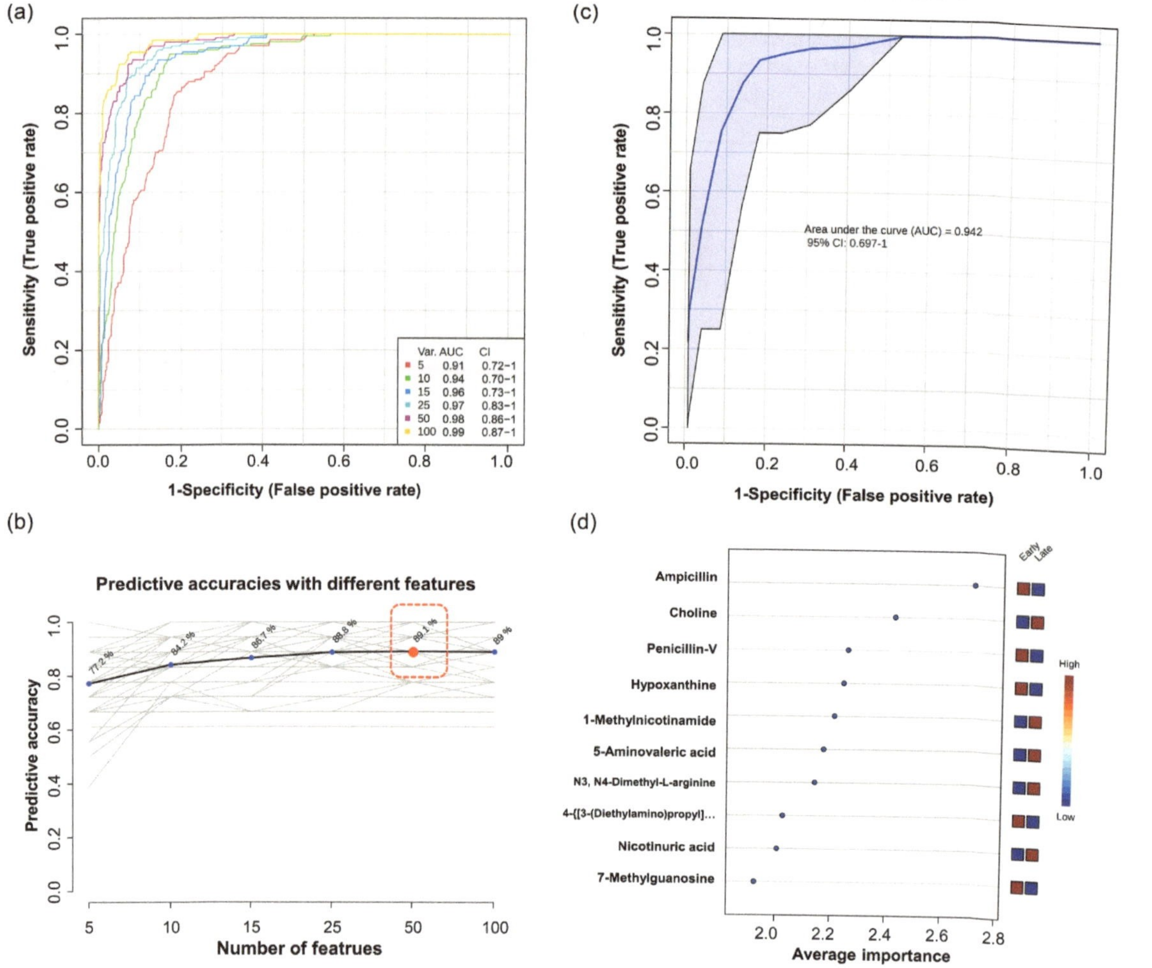

Figure 4. Identification and prediction of key markers between Early group (n = 22) and Late group (n = 12) using multivariate ROC curve-based exploratory analysis. (**a**) Overview of all ROC curves from six distinct predictive models, highlighting their respective AUC values and CI; (**b**) A chart depicting the predictive performance of each of the six models, with the highest accuracy indicated by a red dot for 50−feature panel of model 5; (**c**) The ROC curve specific to the chosen model 2; (**d**) A list of the top 10 significant metabolites, ranked by their average importance of being selected during cross−validation.

4. Discussion

In this study, we analyzed the differences in urinary metabolites between early− and late−stage preterm infants, examining the relationship with maternal transfer of metabolites. Furthermore, we restructured the sample groups to screen preterm infant urine metabolites from various perspectives, focusing on the effects of breastfeeding as a primary mechanism of substance transfer. The analysis revealed that drugs derived from mothers were most distinctly detected in the urine of early−stage preterm infants. Subsequently, significant metabolites related to physiological mechanisms, such as steroid hormones, amino acids, and nucleic acids, were identified. These findings demonstrate that a variety of metabolites are definitively transferred from the mother to the infant post−birth, with some exogenous substances circulating at high concentrations in the infant's body and being substantially excreted between 1 and 3 days post−birth. Furthermore, as a result of the analysis regrouping the sample into an HM group and an FM group, upregulation of some essential amino acids and related metabolites was identified in the urine of the HM group. This underscores the positive impact of breastfeeding on essential amino acid metabolism. This study provides a non−invasive, fundamental approach to understanding the biochemical characteristics in preterm infants, underscoring that the identified alterations in metabolites serve as indirect markers for specific molecular mechanisms. These results are expected to offer novel insights into the multidimensional understanding of the physiological mechanisms in preterm infants and the factors influencing their development and health.

4.1. Investigating the Diverse Physiological Mechanisms and Drug Metabolism in Preterm Infant Urinary Metabolites Using Multivariate Analysis

Initially, this study determined that the metabolite levels in the Early group indicated considerable variability. This substantial deviation in quantitative values is primarily attributed to the physiological immaturity in preterm infants, manifesting in underdeveloped respiratory [19] and hepatic functions [20,21], among other systems. Reflecting these differences, multivariate analysis results demonstrated specific patterns influenced by multifaceted quantitative values. The PCA plot, while showing some overlap between the Early and Late groups, indicated a high similarity within some samples of the same group, showing a distinction between the two groups overall. This pattern suggests that the variability in the Early group is driven by biases from certain variables. Subsequently, the PLS−DA plot emphasized distinctions between the two groups, though minor sub−clusters within each group were observed. These results imply a significant role of certain variables over time post−birth, underscoring the necessity for further analysis and identification of these variables. A similar pattern was also observed in the heatmap analysis. The formation of small−scale clusters within groups, particularly in the Early group, reaffirmed the hypothesis of considerable variation in the quantitative values of specific variables. This finding highlights the complexity of physiological processes in preterm infants and the need for comprehensive multi−variable analysis to better understand these mechanisms. This study aimed to identify key variables contributing to the multidimensional characteristics of the metabolome in our samples. Our approach involved constructing various predictive models and conducting quantitative comparative analyses. Initially, the PLS−DA model highlighted ampicillin as the most significant variable based on its high VIP score and average importance in multivariate ROC curve analysis. Notably, ampicillin has been extensively reported in various studies, including its use in infant fever [22], pharmacokinetics [23,24], and preterm infants [25,26]. Additionally, penicillin−V was also identified as a significant metabolite, exhibiting a high VIP score and average importance. While penicillin antibiotics are known for their safety and low toxicity in neonates, the severity of the use of empirical antibiotics in preterm infants has also been reported [27,28]. The detection of these drugs in high concentrations in urine and systemic circulation suggests a significant association between material transfer between preterm infants and mothers. Particularly, these findings also suggest the influence of external factors such as maternal nutritional status and treatment methods on the metabolism of preterm infants. However, future studies must recognize the potential impact of these substances as significant confounding factors in experiments. This underscores the importance of employing refined methodologies to comprehend intricate interactions in neonatal metabolism.

4.2. Identification of Differential Metabolites in Preterm Infants' Urine over Time via Differential Expression Analysis

Subsequently, we identified DEMs in preterm infants' urine to investigate by comparing quantitative alterations post−birth. To select significant DEMs, we considered VIP score values and results from univariate and multivariate ROC curve analyses. The compound−related functionally significant DEMs were investigated as a priority. As a result of cross−validating various analysis methods with high reliability, we confirmed specific physiological mechanisms post−birth, particularly in drugs, hormones, nucleic acids, and amino acid metabolism. Initially, in the Early group, ampicillin showed the highest log2(FC), identified as the most statistically significant variable in multivariate analyses. Different drugs like penicillin−V, amoxicillin, and lidocaine were also upregulated in the Early group. Considering the compromised immunity in pregnant women and the necessity for drugs in childbirth, drugs such as amoxicillin and ampicillin, classified as penicillin−V, along with lidocaine, are known to be safe and have low toxicity for fetuses [29,30]. Our results show that these exogenous metabolites were directly transferred in high concentrations to preterm infants, being excreted in significant amounts after circulating within 1–3 days. However, carefully administering these drugs is necessary, considering the immature hepatic metabolism in preterm infants [31]. Next, significant biological mechanisms involving steroid hormones, purine and pyrimidine metabolism, and amino acid metabolism were also identified. First, DEM−related steroid hormones like 20β−dihydrocortisol, 17α−hydroxyprogesterone, and pregnenolone indicate which specific hormones play important roles in development. Steroid hormones function as essential metabolites in various physiological processes, including development [32,33] and metabolism [34]. Previous research has reported quantitative hormonal alterations in the urine of premature infants during the transition from the types required for intrauterine and independent life [34]. Among these, 20β−dihydrocortisol and 17α−Hydroxyprogesterone were significantly identified in univariate ROC curve analysis and downregulated in the Early group. Research on 20β−dihydrocortisol, a post−metabolic product of cortisol, was reported as a biomarker for Cushing's syndrome through urine concentration measurements [35]. Additionally, the investigation extends to 17α−Hydroxyprogesterone, a steroid hormone involved in adrenal biosynthesis, transitioning from cholesterol to cortisol [36]. The differential levels of cortisol−related metabolites in preterm infants, depending on the gestational period, underscore their significance in assessing health status [37]. Other upregulated hormone−related substances include pregnenolone as a precursor to steroid hormones [38], 5α−Pregnan−3,20−dione as a neuroactive steroid synthesized from progesterone during fetal development [39], and estriol as a weak estrogen involved in excretion [40]. These differential expressions indirectly provide significant insights into how maternal hormonal states affect metabolic adaptation in preterm infants.

Likewise, some neurotransmitters, similar to hormones, were also differentially expressed in the urine of premature infants. Notably, choline, a precursor to the neurotransmitter acetylcholine and vital for brain differentiation and function, exhibited downregulation in the Early group [41,42]. Choline was identified as a significant metabolite based on its high VIP score and average importance in multivariate ROC curve analysis. Previous research has shown higher concentrations of maternal choline during pregnancy and urinary concentration, with newborns displaying high plasma free choline levels initially, which decrease within the first week [43]. Coupled with this, thiamine was the most significantly downregulated metabolite. Thiamine is a water−soluble vitamin B1 and plays a direct role in the synthesis and release of acetylcholine, with its deficiency linked to reduced acetylcholine levels [44,45]. Considering these, the downregulation of two significant metabolites in the Early group suggests potential negative markers for cognitive and behavioral development in preterm infants. Conversely, other DEMs such as taurine, S−adenosylmethionine, and L−pyroglutamic acid showed significant VIP scores and upregulation trends in the Early group. Taurine, known to function as a neurotransmitter or modulator, has been indicated as a marker for muscle damage from severe exercise [46], while L−pyroglutamic acid is closely related to the major neurotransmitter glutamate and found in high concentrations in urine [47]. Moreover, the increase in neurotransmitter−related metabolite concentrations suggests that these DEMs might be markers of brain diseases or neurodevelopmental disorders for use in urine−based studies of preterm infants. From another perspective, some purine and pyrimidine metabolism− and amino acid−related metabolites were also identified as significant DEMs, participating in fundamental biological mechanisms due to their role in DNA composition, genetic factors, cellular structure, and energy production [48–50]. In this study, we observed significant upregulation of hypoxanthine and guanine in the Early group of infants with significant VIP scores and average importance. Hypoxanthine is a primary breakdown product of ATP, and guanine is a representative nucleic acid of DNA building blocks [51]. According to previous research, high urinary hypoxanthine levels can be related to respiratory disorders and deficiencies in hypoxanthine–guanine phosphoribosyltransferase to acute renal failure in infants [52,53]. Additionally, 7−methylguanosine, the modified purine nucleoside, showed significant VIP scores, associated with ischemic diseases and identified as a biomarker in cancer studies [54,55]. In brief, these upregulations suggest significant metabolic mechanisms related to kidney and respiratory development. Furthermore, amino acids, essential for protein construction and energy in preterm infants, were also notably identified [56]. Among the notable DEMs, acetyl−L−carnitine associated with catabolic and anabolic metabolism in the brain as an endogenous intermediate [57], N−acetyl−L−tyrosine administered for stability and enhanced solubility in premature infants during the first postnatal week [58], and other energy metabolism substances such as hexanoylcarnitine [59], as well as fundamental amino acids such as L−tyrosine and L−phenylalanine, have been identified. Amino acids are multifunctional and actively under investigation from various perspectives. Consequently, imbalanced urinary concentrations of amino acids and associated metabolites are indicative of their influence on maternal conditions and neonatal metabolic stress, encouraging deeper exploration associated with these results.

4.3. *Efficacy of Human Milk in Premature Infants from a Substance Transfer Perspective and Essential Amino Acids*

In this study, we aimed to explore the additional metabolomic profile using preterm infant urine, focusing on another perspective. To investigate this, we regrouped our sample cohorts based on the intake of HM versus FM. Through additional analysis, we revealed that human milk consumption plays a significant role in the supply of essential amino acids. The PCA plot did not show a distinct separation between the HM and FM groups, likely due to the high concentration of specific substances like ampicillin (Figure S1). Nevertheless, among the 12 identified DEMs, isoleucine, imidazolelactic acid, and DL−β−leucine exhibited an intriguing upregulation in the HM group (Figure 5, Table S7). Isoleucine, an essential amino acid involved in the tricarboxylic acid cycle, serves as a fundamental factor in energy metabolism, as well as supplying acetyl−CoA, which is a crucial intermediary in neurotransmitter and steroid synthesis [60,61]. The detection of imidazolelactic acid in urine, which is produced through the breakdown of histidine by an alternative pathway in the absence of histidase, has been reported in several studies [62,63]. Histidine is particularly essential in infancy, and underscored in conjunction with L−tyrosine, previously identified as significant. This characteristic arises due to the immature enzymatic systems in newborns, highlighting their critical metabolic role in preterm infants [64,65]. DL−β−leucine is a less abundant β−amino acid compared to its α−analogues but exists in nature both in free-form and peptide−bound states. Although little research has been conducted, previous research for gestational diabetes mellitus using LC−MS has identified it, suggesting its potential specific functions as analogs of leucine, one of the essential amino acids, in preterm infants [66]. In summary, the upregulation of certain essential amino acids in the HM group indicates a significant impact on metabolism immediately after preterm birth. By focusing on breastfeeding influence in terms of direct nutrient delivery, we investigated its impact. This suggests that human milk feeding, compared to formula feeding, might be more effective in delivering these amino acids that are crucial for development. Certainly, further research is required for a full understanding of the biological meaning, but the accumulation of such data may provide insights into the relationship and molecular mechanisms between the mother and preterm infant in future studies.

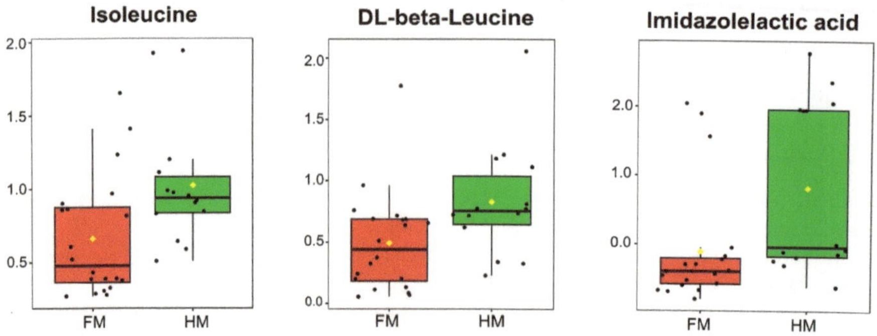

Figure 5. Box–whisker plots of three upregulated metabolites in HM group ($n = 14$) compared to the FM ($n = 20$). Boxplots of relative concentrations for selected DEMs between FM (red) and HM (green) groups. Black dots denote the concentration levels of the chosen feature across all samples. Horizontal red lines mark the optimal cutoff, while the average concentration for each group is symbolized by yellow diamonds.

5. Conclusions

In this study, we used metabolomics to analyze differences in urine metabolites between early− and late−stage preterm infants. As a result, drugs derived from mothers were most prominently detected in the urine of early−stage preterm infants, confirming substantial excretion within 1–3 days post−birth. Various key metabolites, including hormones such as pregnenolone and 20β−dihydrocortisol, purine and pyrimidine metabolism-related metabolites like hypoxanthine and guanine, neurotransmitters including choline and L−pyroglutamic acid, and amino acids such as acetyl−L−carnitine and L−tyrosine, were identified through good predictive models and differential expression analysis, demonstrating statistically validated suitability and performance. Our study used LC−MS to analyze urine samples from preterm infants, similar to previous research, but our approach revealed unique physiological mechanisms, demonstrating that LC−MS offers essential insights into the intricate metabolic processes of preterm infants, despite using similar samples and controls. Additionally, this study suggests the potential involvement of substances in preterm infant development and highlights the positive impact of breastfeeding on essential amino acid metabolism. The non−invasive sampling and high sensitivity of this research indirectly indicate the association between mothers and premature infants, providing insights into their biochemical characteristics and physiological mechanisms. Furthermore, the identification of significant metabolites serving as indirect markers for specific molecular mechanisms contributes to the understanding of preterm infant physiology and potential biomarkers for specific clinical studies.

Supplementary Materials: The following supporting information can be downloaded at https://www.mdpi.com/article/10.3390/nu16030411/s1, Figure S1: PCA plot for HM and FM group; Table S1: List of 284 identified metabolites; Table S2: Top 30 VIP score list in PLS−DA model; Table S3: List of 156 identified DEMs in Early/Late group; Table S4: Th selected metabolite list by LASSO algorithm; Table S5: Univariate ROC curve analysis results; Table S6: Multivariate ROC curve analysis results; Table S7: List of 12 identified DEMs in HM/FM group.

Author Contributions: Conceptualization, J.-H.M., J.S., W.-H.H., J.-M.P., J.K., H.K. and N.M.K.; methodology, J.-H.M. and S.C.; software, J.-H.M., J.S. and S.C.; validation, J.-H.M. and S.C.; formal analysis, J.-H.M.; investigation, J.-H.M.; resources, J.-H.M.; data curation, J.-H.M.; writing—original draft preparation, J.-H.M.; writing—review and editing, J.-H.M.; visualization, J.-H.M. and J.S.; supervision, J.-H.M., J.S. and W.-H.H.; project administration, J.S., W.-H.H., J.-M.P. and N.M.K.; funding acquisition, J.S., W.-H.H., J.-M.P., J.K., H.K. and N.M.K. All authors have read and agreed to the published version of the manuscript.

Funding: This work was supported by the National Research Foundation of Korea (NRF) grant funded by the Korean government (MSIT). (No. 2020R1A2C1005082); Korea-EU (ERC) Joint Research Program funded by the MSIT through the NRF of Korea (No. RS-2023-00265366).

Institutional Review Board Statement: All subjects gave their informed consent for inclusion before they participated in the study. The study was conducted in accordance with the Declaration of Helsinki and approved by the Institutional Review Board of Soonchunhyang University Hospital Cheonan (SCHCA 2020-08-034).

Informed Consent Statement: Not applicable.

Data Availability Statement: Data are contained within the article or Supplementary Materials.

Conflicts of Interest: The authors declare no conflict of interest. S.C. and J.-M.P. are employees of Basil Biotech. This paper reflects the views of the scientists and not the company.

References

1. Abu-Saad, K.; Fraser, D. Maternal Nutrition and Birth Outcomes. *Epidemiol. Rev.* **2010**, *32*, 5–25. [CrossRef]
2. Dewey, K.G. Impact of Breastfeeding on Maternal Nutritional Status. In *Protecting Infants through Human Milk*; Springer: Boston, MA, USA, 2004; pp. 91–100.
3. Alamri, S.H.; Abdeen, G.N. Maternal Nutritional Status and Pregnancy Outcomes Post-bariatric Surgery. *Obes. Surg.* **2022**, *32*, 1325–1340. [CrossRef]
4. Khalili Doroodzani, A.; Dobaradaran, S.; Zarei, S.; Raeisi, A.; Mahmoodi, M.; Rahmani, E.; Nabipour, I.; Saeedi, R.; Mahmudpour, M.; Akbarzadeh, S.; et al. Maternal and fetal exposure to metal (loid)s, maternal nutrition status, and impact on prenatal growth in an energy rich zone and an urban area along the Persian Gulf. *Environ. Pollut.* **2022**, *309*, 119779. [CrossRef]
5. Zaidi, A.Z.; Moore, S.E.; Okala, S.G. Impact of Maternal Nutritional Supplementation during Pregnancy and Lactation on the Infant Gut or Breastmilk Microbiota: A Systematic Review. *Nutrients* **2021**, *13*, 1137. [CrossRef] [PubMed]
6. Greer, F.R.; Sicherer, S.H.; Burks, A.W.; Committee on Nutrition; Section on Allergy and Immunology; Abrams, S.A.; Fuchs, G.J., III; Kim, J.H.; Lindsey, C.W. The Effects of Early Nutritional Interventions on the Development of Atopic Disease in Infants and Children: The Role of Maternal Dietary Restriction, Breastfeeding, Hydrolyzed Formulas, and Timing of Introduction of Allergenic Complementary Foods. *Pediatrics* **2019**, *143*, e20190281. [CrossRef] [PubMed]
7. Jobgen, W.S.; Fried, S.K.; Fu, W.J.; Meininger, C.J.; Wu, G. Regulatory role for the arginine-nitric oxide pathway in metabolism of energy substrates. *J. Nutr. Biochem.* **2006**, *17*, 571–588. [CrossRef] [PubMed]
8. Wu, G.; Bazer, F.W.; Datta, S.; Johnson, G.A.; Li, P.; Satterfield, M.C.; Spencer, T.E. Proline metabolism in the conceptus: Implications for fetal growth and development. *Amino Acids* **2008**, *35*, 691–702. [CrossRef]
9. Haggarty, P. Fatty acid supply to the human fetus. *Annu. Rev. Nutr.* **2010**, *30*, 237–255. [CrossRef] [PubMed]
10. Wu, G. Amino acids: Metabolism, functions, and nutrition. *Amino Acids* **2009**, *37*, 1–17. [CrossRef]
11. Wu, G.; Bazer, F.W.; Burghardt, R.C.; Johnson, G.A.; Kim, S.W.; Knabe, D.A.; Li, P.; Li, X.; McKnight, J.R.; Satterfield, M.C.; et al. Proline and hydroxyproline metabolism: Implications for animal and human nutrition. *Amino Acids* **2011**, *40*, 1053–1063. [CrossRef]
12. Moltu, S.J.; Sachse, D.; Blakstad, E.W.; Strømmen, K.; Nakstad, B.; Almaas, A.N.; Westerberg, A.C.; Rønnestad, A.; Brække, K.; Veierød, M.B.; et al. Urinary Metabolite Profiles in Premature Infants Show Early Postnatal Metabolic Adaptation and Maturation. *Nutrients* **2014**, *6*, 1913–1930. [CrossRef]
13. Emwas, A.H. The strengths and weaknesses of NMR spectroscopy and mass spectrometry with particular focus on metabolomics research. *Methods Mol. Biol.* **2015**, *1277*, 161–193. [CrossRef]
14. West, C.R.; Harding, J.E.; Williams, C.E.; Gunning, M.I.; Battin, M.R. Quantitative electroencephalographic patterns in normal preterm infants over the first week after birth. *Early Hum. Dev.* **2006**, *82*, 43–51. [CrossRef]
15. Knobel-Dail, R.B.; Tanaka, D.T.; Holditch-Davis, D.; White, J. Perfusion Index in Very Low Birth Weight Premature Infants During Their First 2 Weeks of Life. *Biol. Res. Nurs.* **2016**, *19*, 45–52. [CrossRef] [PubMed]
16. Fajardo, C.; Alshaikh, B.; Harabor, A. Prolonged use of antibiotics after birth is associated with increased morbidity in preterm infants with negative cultures. *J. Matern. Fetal Neonatal Med.* **2019**, *32*, 4060–4066. [CrossRef] [PubMed]
17. De Curtis, M.; Rigo, J. The nutrition of preterm infants. *Early Hum. Dev.* **2012**, *88*, S5–S7. [CrossRef] [PubMed]
18. Fiehn, O.; Robertson, D.; Griffin, J.; van der Werf, M.; Nikolau, B.; Morrison, N.; Sumner, L.W.; Goodacre, R.; Hardy, N.W.; Taylor, C.; et al. The metabolomics standards initiative (MSI). *Metabolomics* **2007**, *3*, 175–178. [CrossRef]
19. Smith, L.J.; McKay, K.O.; van Asperen, P.P.; Selvadurai, H.; Fitzgerald, D.A. Normal Development of the Lung and Premature Birth. *Paediatr. Respir. Rev.* **2010**, *11*, 135–142. [CrossRef] [PubMed]
20. Grijalva, J.; Vakili, K. Neonatal liver physiology. *Semin. Pediatr. Surg.* **2013**, *22*, 185–189. [CrossRef] [PubMed]
21. Machado, J.S.; Ferreira, T.S.; Lima, R.C.G.; Vieira, V.C.; Medeiros, D.S. Premature birth: Topics in physiology and pharmacological characteristics. *Rev. Assoc. Med. Bras.* **2021**, *67*, 150–155. [CrossRef] [PubMed]
22. Brown, J.C.; Burns, J.L.; Cummings, P. Ampicillin Use in Infant Fever: A Systematic Review. *Arch. Pediatr. Adolesc. Med.* **2002**, *156*, 27–32. [CrossRef]

23. Yoshioka, H.; Takimoto, M.; Riley, H.D., Jr. Pharmacokinetics of Ampicillin in the Newborn Infant. *J. Infect. Dis.* **1974**, *129*, 461–464. [CrossRef]
24. Ginsburg, C.M.; McCracken, G.H., Jr.; Thomas, M.L.; Clahsen, J. Comparative Pharmacokinetics of Amoxicillin and Ampicillin in Infants and Children. *Pediatrics* **1979**, *64*, 627–631. [CrossRef]
25. Axline, S.G.; Yaffe, S.J.; Simon, H.J. Clinical Pharmacology of Antimicrobials in Premature Infants: II. Ampicillin, Methicillin, Oxacillin, Neomycin, and Colistin. *Pediatrics* **1967**, *39*, 97–107. [CrossRef] [PubMed]
26. Le, J.; Greenberg, R.G.; Yoo, Y.; Clark, R.H.; Benjamin Jr, D.K.; Zimmerman, K.O.; Cohen-Wolkowiez, M.; Wade, K.C.; Best Pharmaceuticals for Children Act—Pediatric Trials Network Steering Committee. Ampicillin dosing in premature infants for early-onset sepsis: Exposure-driven efficacy, safety, and stewardship. *J. Perinatol.* **2022**, *42*, 959–964. [CrossRef] [PubMed]
27. Barker, C.I.; Germovsek, E.; Sharland, M. What do I need to know about penicillin antibiotics? *Arch. Dis. Child. Educ. Pract. Ed.* **2017**, *102*, 44–50. [CrossRef] [PubMed]
28. Kuppala, V.S.; Meinzen-Derr, J.; Morrow, A.L.; Schibler, K.R. Prolonged Initial Empirical Antibiotic Treatment is Associated with Adverse Outcomes in Premature Infants. *J. Pediatr.* **2011**, *159*, 720–725. [CrossRef] [PubMed]
29. Dashe, J.S.; Gilstrap, L.C. Antibiotic Use in Pregnancy. *Obstet. Gynecol. Clin. N. Am.* **1997**, *24*, 617–629. [CrossRef] [PubMed]
30. Favero, V.; Bacci, C.; Volpato, A.; Bandiera, M.; Favero, L.; Zanette, G. Pregnancy and dentistry: A literature review on risk management during dental surgical procedures. *Dent. J.* **2021**, *9*, 46. [CrossRef] [PubMed]
31. Hottinger, A. Liver-immaturity and treatment of kernicterus of the premature baby. *Indian. J. Pediatr.* **1958**, *25*, 184–200. [CrossRef]
32. McEwen, B.S. Steroid Hormones: Effect on Brain Development and Function. *Horm. Res.* **2008**, *37*, 1–10. [CrossRef] [PubMed]
33. Greaves, R.F.; Pitkin, J.; Ho, C.S.; Baglin, J.; Hunt, R.W.; Zacharin, M.R. Hormone Modeling in Preterm Neonates: Establishment of Pituitary and Steroid Hormone Reference Intervals. *J. Clin. Endocrinol. Metab.* **2015**, *100*, 1097–1103. [CrossRef]
34. Mitchell, F.L.; Shackleton, C.H.L. The Investigation of Steroid Metabolism in Early Infancy. In *Advances in Clinical Chemistry*; Bodansky, O., Stewart, C.P., Eds.; Elsevier: Amsterdam, The Netherlands, 1969; Volume 12, pp. 141–215.
35. Tomlinson, J.W.; Stewart, P.M. Cortisol metabolism and the role of 11β-hydroxysteroid dehydrogenase. *Best. Pract. Res. Clin. Endocrinol. Metab.* **2001**, *15*, 61–78. [CrossRef] [PubMed]
36. Honour, J.W. 17-Hydroxyprogesterone in children, adolescents and adults. *Ann. Clin. Biochem.* **2014**, *51*, 424–440. [CrossRef] [PubMed]
37. Ng, S.M.; Ogundiya, A.; Didi, M.; Turner, M.A. Adrenal function of extremely premature infants in the first 5 days after birth. *J. Pediatr. Endocrinol. Metab.* **2019**, *32*, 363–367. [CrossRef] [PubMed]
38. Prasad, V.; Vegesna, S.R.; Welch, M.; Lieberman, S. Precursors of the neurosteroids. *Proc. Natl. Acad. Sci. USA* **1994**, *91*, 3220–3223. [CrossRef] [PubMed]
39. Wang, T.; Yao, J.; Chen, S.; Mao, Z.; Brinton, R.D. Allopregnanolone Reverses Bioenergetic Deficits in Female Triple Transgenic Alzheimer's Mouse Model. *Neurotherapeutics* **2020**, *17*, 178–188. [CrossRef]
40. Kuhl, H. Pharmacology of estrogens and progestogens: Influence of different routes of administration. *Climacteric* **2005**, *8*, 3–63. [CrossRef]
41. Bernhard, W.; Poets, C.F.; Franz, A.R. Choline and choline-related nutrients in regular and preterm infant growth. *Eur. J. Nutr.* **2019**, *58*, 931–945. [CrossRef]
42. Bernhard, W.; Full, A.; Arand, J.; Maas, C.; Poets, C.F.; Franz, A.R. Choline supply of preterm infants: Assessment of dietary intake and pathophysiological considerations. *Eur. J. Nutr.* **2013**, *52*, 1269–1278. [CrossRef]
43. McMahon, K.E.; Farrell, P.M. Measurement of free choline concentrations in maternal and neonatal blood by micropyrolysis gas chromatography. *Clin. Chim. Acta* **1985**, *149*, 1–12. [CrossRef]
44. Eder, L.; Hirt, L.; Dunant, Y. Possible involvement of thiamine in acetylcholine release. *Nature* **1976**, *264*, 186–188. [CrossRef]
45. Speeg, K., Jr.; Chen, D.; McCandless, D.; Schenker, S. Cerebral acetylcholine in thiamine deficiency. *Proc. Soc. Exp. Biol. Med.* **1970**, *134*, 1005–1009. [CrossRef] [PubMed]
46. Cuisinier, C.; Ward, R.J.; Francaux, M.; Sturbois, X.; de Witte, P. Changes in plasma and urinary taurine and amino acids in runners immediately and 24 h after a marathon. *Amino Acids* **2001**, *20*, 13–23. [CrossRef]
47. Silva, A.R.; Silva, C.G.; Ruschel, C.; Helegda, C.; Wyse, A.T.; Wannmacher, C.M.; Wajner, M.; Dutra-Filho, C.S. L-pyroglutamic acid inhibits energy production and lipid synthesis in cerebral cortex of young rats in vitro. *Neurochem. Res.* **2001**, *26*, 1277–1283. [CrossRef] [PubMed]
48. Kamatani, N.; Jinnah, H.A.; Hennekam, R.C.M.; van Kuilenburg, A.B.P. 6—Purine and Pyrimidine Metabolism. In *Emery and Rimoin's Principles and Practice of Medical Genetics and Genomics*, 7th ed.; Pyeritz, R.E., Korf, B.R., Grody, W.W., Eds.; Academic Press: Cambridge, MA, USA, 2021; pp. 183–234. [CrossRef]
49. Van den Berghe, G.; Vincent, M.F. Disorders of Purine and Pyrimidine Metabolism. In *Inborn Metabolic Diseases: Diagnosis and Treatment*; Fernandes, J., Saudubray, J.-M., Van den Berghe, G., Tada, K., Buist, N.R.M., Eds.; Springer: Berlin/Heidelberg, Germany, 1995; pp. 289–302. [CrossRef]
50. Perrone, S.; Laschi, E.; De Bernardo, G.; Giordano, M.; Vanacore, F.; Tassini, M.; Calderisi, M.; Toni, A.L.; Buonocore, G.; Longini, M. Newborn metabolomic profile mirrors that of mother in pregnancy. *Med. Hypotheses* **2020**, *137*, 109543. [CrossRef] [PubMed]
51. Holden, M.S.; Hopper, A.; Slater, L.; Asmerom, Y.; Esiaba, I.; Boskovic, D.S.; Angeles, D.M. Urinary Hypoxanthine as a Measure of Increased ATP Utilization in Late Preterm Infants. *ICAN Infant Child Adolesc. Nutr.* **2014**, *6*, 240–249. [CrossRef]

52. Wingen, A.M.; Löffler, W.; Waldherr, R.; Schärer, K. Acute renal failure in an infant with partial deficiency of hypoxanthine-guanine phosphoribosyltransferase. *Proc. Eur. Dial. Transplant. Assoc. Eur. Ren. Assoc.* **1985**, *21*, 751–755.
53. Holland, P.C.; Dillon, M.J.; Pincott, J.; Simmonds, H.A.; Barratt, T.M. Hypoxanthine guanine phosphoribosyl transferase deficiency presenting with gout and renal failure in infancy. *Arch. Dis. Child.* **1983**, *58*, 831–833. [CrossRef]
54. Xiao, Y.; Yang, J.; Yang, M.; Len, J.; Yu, Y. Comprehensive analysis of 7-methylguanosine and immune microenvironment characteristics in clear cell renal cell carcinomas. *Front. Genet.* **2022**, *13*, 866819. [CrossRef]
55. Raćkowska, E.; Bobrowska-Korczak, B.; Giebułtowicz, J. Development and validation of a rapid LC–MS/MS method for determination of methylated nucleosides and nucleobases in urine. *J. Chromatogr. B* **2019**, *1128*, 121775. [CrossRef]
56. Embleton, N.D. Optimal protein and energy intakes in preterm infants. *Early Human. Dev.* **2007**, *83*, 831–837. [CrossRef]
57. Dolezal, V.; Tucek, S. Utilization of citrate, acetylcarnitine, acetate, pyruvate and glucose for the synthesis of acetylcholine in rat brain slices. *J. Neurochem.* **1981**, *36*, 1323–1330. [CrossRef]
58. Van Goudoever, J.B.; Sulkers, E.J.; Timmerman, M.; Huijmans, J.G.M.; Langer, K.; Carnielli, V.P.; Sauer, P.J.J. Amino Acid Solutions for Premature Neonates During the First Week of Life: The Role of N-Acetyl-L-Cysteine and N-Acetyl-L-Tyrosine. *J. Parenter. Enter. Nutr.* **1994**, *18*, 404–408. [CrossRef]
59. du Toit, W.L.; Kruger, R.; Gafane-Matemane, L.F.; Schutte, A.E.; Louw, R.; Mels, C.M. Urinary metabolomics profiling by cardiovascular risk factors in young adults: The African Prospective study on Early Detection and Identification of Cardiovascular disease and Hypertension study. *J. Hypertens.* **2022**, *40*, 1545–1555. [CrossRef]
60. Johansen, M.L.; Bak, L.K.; Schousboe, A.; Iversen, P.; Sørensen, M.; Keiding, S.; Vilstrup, H.; Gjedde, A.; Ott, P.; Waagepetersen, H.S. The metabolic role of isoleucine in detoxification of ammonia in cultured mouse neurons and astrocytes. *Neurochem. Int.* **2007**, *50*, 1042–1051. [CrossRef]
61. Gu, C.; Mao, X.; Chen, D.; Yu, B.; Yang, Q. Isoleucine plays an important role for maintaining immune function. *Curr. Protein Pept. Sci.* **2019**, *20*, 644–651. [CrossRef] [PubMed]
62. Cain, A.; Holton, J. Histidinaemia: A child and his family. *Arch. Dis. Child.* **1968**, *43*, 62. [CrossRef] [PubMed]
63. Kappelman, M.; Thomas, G.H.; Howell, R.R. Histidinemia in a Negro child. *Am. J. Dis. Child.* **1971**, *122*, 212–214. [CrossRef] [PubMed]
64. Pencharz, P.B.; Ball, R.O. Amino acid requirements of infants and children. *Protein Energy Requir. Infancy Child.* **2006**, *58*, 109–119.
65. Heird, W.C. Amino acids in pediatric and neonatal nutrition. *Curr. Opin. Clin. Nutr. Metab. Care* **1998**, *1*, 73–78. [CrossRef] [PubMed]
66. Prabhu, M.D.; Tonannavar Yenagi, J.; Kamat, V.; Tonannavar, J. XRD structure and vibrational analysis of DL-β-Leucine, as aided by DFT tetramer model and characterized by NBO, AIM and NCI calculations. *J. Mol. Struct.* **2020**, *1218*, 128495. [CrossRef]

nutrients

MDPI

Review

The Influence of Early Nutrition on Neurodevelopmental Outcomes in Preterm Infants

Rita C. Silveira, Andrea L. Corso and Renato S. Procianoy *

Department of Pediatrics, Newborn Section, Hospital de Clínicas de Porto Alegre, Universidade Federal do Rio Grande do Sul, Porto Alegre 3452925, RS, Brazil
* Correspondence: rprocianoy@gmail.com

Abstract: Premature infants, given their limited reserves, heightened energy requirements, and susceptibility to nutritional deficits, require specialized care. Aim: To examine the complex interplay between nutrition and neurodevelopment in premature infants, underscoring the critical need for tailored nutritional approaches to support optimal brain growth and function. Data sources: PubMed and MeSH and keywords: preterm, early nutrition, macronutrients, micronutrients, human milk, human milk oligosaccharides, probiotics AND neurodevelopment or neurodevelopment outcomes. Recent articles were selected according to the authors' judgment of their relevance. Specific nutrients, including macro (amino acids, glucose, and lipids) and micronutrients, play an important role in promoting neurodevelopment. Early and aggressive nutrition has shown promise, as has recognizing glucose as the primary energy source for the developing brain. Long-chain polyunsaturated fatty acids, such as DHA, contribute to brain maturation, while the benefits of human milk, human milk oligosaccharides, and probiotics on neurodevelopment via the gut-brain axis are explored. This intricate interplay between the gut microbiota and the central nervous system highlights human milk oligosaccharides' role in early brain maturation. Conclusions: Individualized nutritional approaches and comprehensive nutrient strategies are paramount to enhancing neurodevelopment in premature infants, underscoring human milk's potential as the gold standard of nutrition for preterm infants.

Keywords: preterm; early nutrition; macronutrients; micronutrients; human milk; human milk oligosaccharides; probiotics

1. Introduction

Preterm infants constitute a diverse population with unique nutritional requirements that necessitate individualized approaches tailored to their clinical status and degree of prematurity. The overarching goal is to enhance neurodevelopmental outcomes, all while acknowledging the nutritional challenges posed by preterm birth. The brain, being the most metabolically active organ in premature infants, demands a substantial supply of nutrients for optimal growth and functional development. While all nutrients play vital roles, some exert a more pronounced impact, particularly during the critical period spanning from 24 to 52 weeks post-conceptional age. This timeframe represents a pivotal window for neurodevelopment, encompassing the development of essential structures such as white and gray matter, cell replication, neurogenesis, neuronal and cerebral white matter differentiation, cell migration, myelination, and synaptogenesis, among other intricate processes [1–3].

The abrupt cessation of placental nutrient delivery places preterm infants at a heightened risk for restricted postnatal growth. These infants are born with limited reserves of essential nutrients, exhibit compromised thermoregulation, and have elevated energy requirements. The primary challenge lies in identifying the key nutrients crucial for optimal brain development, considering factors like the neonate's gestational age, underlying morbidities, enteral versus parenteral feeding capacities, and growth requirements. The early initiation of aggressive enteral feeding alongside appropriate parenteral support holds the potential to enhance the growth and overall development of very low-birth-weight and extremely premature infants. Conversely, excessive early nutritional support in preterm infants may lead to alterations in body composition and subsequently increase the risk of obesity and chronic non-communicable diseases later in life [1,4,5].

Recently, the concept of the microbiota-gut-brain axis has gained prominence, shedding light on the intricate interplay between the intestinal microbiota and the central nervous system. Accumulating evidence underscores the significance of the gut microbiome in this bidirectional communication system, constituting a complex network that modulates immune, gastrointestinal, and central nervous system functions [6,7].

The term "gut microbiome" encompasses the complex ecosystem of bacteria inhabiting the intestine, including their genes, proteins, and metabolites. Investigations into the interaction between the intestinal microbiota and the central nervous system have unveiled critical developmental windows, especially in the vulnerable preterm population. Manipulating the intestinal microbiota through prebiotic and probiotic supplementation presents a promising avenue for improving neurobehavioral outcomes in preterm infants. For instance, the absence of Bifidobacterium during the first month of life in preterm infants has been associated with delayed neurodevelopment in early childhood. The administration of Bifidobacterium strains may foster optimal neurocognitive development in these vulnerable children [6–8].

Hence, any nutritional strategy designed to mitigate initial weight loss and foster brain growth holds the potential to significantly enhance neurodevelopmental outcomes in preterm infants, particularly those born extremely prematurely, who traditionally face a higher risk of adverse outcomes. This review aims to provide a comprehensive overview of the current understanding of nutritional interventions, from neonatal to post-hospital discharge and follow-up, that can positively influence neurodevelopment. Specific nutrients, such as amino acids and lipids, nutritional supplements, and dietary practices, including breastfeeding and its undeniable benefits, are addressed. In this review, we address neuro-nutrition as a key concept of brain health, where each nutrient has its specific function in the developing brain of preterm infants (Table 1).

Table 1. Neuro-nutrition: The impact of each nutrient on the brain.

Nutrient	Impact	Brain Structure
Energy and protein	Cell multiplication and differentiation, synaptogenesis, growth factors	Cortex, hippocampus, global brain
Iron	Myelin, monoamine synthesis, glial metabolism	White matter and hippocampus
Zinc	DNA synthesis, neurotransmitters	Autonomous nervous system, hippocampus, cerebellum
Copper	Neurotransmitters, glial metabolism, antioxidation	Cerebellum
LC-PUFAS	Synaptogenesis	Retina, cortex
TAURINE/HILL	Neurotransmitters, DNA methylination, myelin	Global region, hippocampus and cerebral white matter

Adapted from [9,10].

2. Data Sources

Data sources: PubMed and MeSH and keywords: preterm, early nutrition, macronutrients, micronutrients, human milk, human milk oligosaccharides, probiotics AND neurodevelopment or neurodevelopment outcomes. Recent articles were selected according to the authors' judgment of their relevance.

3. Amino Acids

The literature extensively details the cognitive benefits of early, aggressive amino acid supplementation. Very low-birth-weight preterm infants who receive higher protein and energy intake during their first week of life exhibit improved IQ scores and reduced developmental delays at 18 months of age. This positive outcome is often accompanied by adequate head-circumference catch-up growth or accelerated head circumference growth. Additionally, very low-birth-weight premature infants with substantial head circumference growth tend to demonstrate better cognitive outcomes [11–13].

Despite the structural and functional immaturity of the intestines, which may hinder enteral nutrition progress, early initiation of enteral feeding is strongly recommended. Cormack and colleagues established a positive correlation between enteral protein intake in the first two weeks of life and cognitive and motor subscale scores on the Bayley III scale at 18 months of corrected age [14].

Evidence indicates that nutritional deficits at 28 days after birth are negatively associated with neurodevelopment at 3 months of age. Furthermore, studies have shown that higher energy and protein intake during the first week after birth are linked to improved neurodevelopment at 18 months of corrected age in premature infants [15,16].

A Cochrane review supports the early provision of amino acids in parenteral nutrition, leading to increased head circumference growth and reduced postnatal growth failure at discharge. The number needed to treat for an additional beneficial outcome is 7 (NNT = 7) [17]. Subgroup analyses have indicated a significant reduction in postnatal growth failure at discharge for preterm infants receiving high amino acid intake in parenteral nutrition (>2 to ≤3 g/kg/d). However, the impact on neurodevelopmental outcomes remains inconclusive [17].

In practice, the recommended target dose of amino acids in parenteral solutions remains under discussion. Suggestions for premature infants range from 3 g/kg/day to 3.5 g/kg/day of amino acids. There is no evidence to support better neurodevelopment with a maximum dose of 4 g/kg/day [18].

Determining the optimal amino acid amount in enteral nutrition for formula-fed low-birth weight preterm infants remains a subject of debate. Protein intake should suffice to achieve normal growth without causing undesirable effects such as acidosis, uremia, or elevated circulating amino acid levels. Simultaneously, it should aim to enhance head circumference growth and subsequently improve developmental outcomes. A Cochrane meta-analysis comparing low amino acid intake (<3 g/kg/d) versus high amino acid intake (3.0 to 4.0 g/kg/d) in formulas found that high protein intake (≥3.0 g/kg/d and <4.0 g/kg/d) accelerated the weight gain of preterm infants, although methodological limitations hindered a conclusive determination of its favorability [19].

Although breast milk is considered the gold standard for nutrition, its protein concentration may not meet the needs of preterm infants due to limitations in enteral volume supply. Therefore, breast milk is often supplemented with proteins and other nutrients [3]. However, determining the optimal concentration of commercially available proteins remains inconclusive despite numerous studies and reviews aiming to enhance nutritional performance for neurodevelopment [2,13,14,20].

Kumar et al., previously reinforced the role of individualized human milk fortification through two methods: target and adjustable fortification. Target fortification involves analyzing the protein content of human milk and adding additives based on the baby's specific nutritional requirements, pre-defined by the care team. In adjustable fortification, protein intake is periodically adjusted based on the child's metabolic response assessed by laboratory tests, making it more suitable for stable premature babies and practical, as it does not require a human milk analysis. Fortification significantly impacts head circumference growth and neurodevelopment [3].

A systematic review encompassing nine studies with a total of 861 premature infants found that a high protein concentration (≥1.4 g/100 mL) in human milk additives derived from bovine milk increased in-hospital weight gain compared to a moderate protein concentration (≥1 g < 1.4 g/100 mL). However, there was insufficient evidence to assess the impact of protein concentration on adverse effects or long-term outcomes like neurodevelopment [20].

Currently, recommendations suggest providing premature babies with a gestational age below 32 weeks at birth with 3.5 to 4.5 g/kg of proteins per day. As enteral supply improves, amino acids in parenteral nutrition should be reduced to ensure that the total protein intake does not exceed 4.5 g/kg per day (combined enteral and parenteral). Limited evidence suggests that a protein-to-calorie ratio of 30 to 40 kcal per gram can maximize protein synthesis and positively impact the neurodevelopment of premature infants [21].

It is worth noting that the type of amino acid offered may be more influential than the quantity in terms of neurocognition. Low taurine levels in the neonatal period of premature infants negatively affect neurological development, highlighting the advantage of breast milk, which contains high levels of taurine, the most abundant free amino acid in human milk, along with glutamate [22]. Taurine's beneficial role primarily involves auditory and visual development and is frequently supplemented in infant formulas [21].

In the developing brain, GABA and glutamate play essential roles in neurotransmission, neuronal migration, dendrite and synapse formation, and neural circuit organization. The abundance of glutamate, relative to other acids, in human milk reinforces its neurocognitive benefits [23].

4. Glucose

Glucose serves as the primary energy source for the brain and nervous system, with the human brain consuming about 20% of the total body's glucose in normal circumstances. In cases of limited glucose availability, the brain can utilize lactate and ketone bodies as alternative energy sources. Lactate sustains brain activity during glucose deprivation and supports several neuronal functions [24].

Owing to their premature birth, preterm infants have inadequate energy reserves, as glycogen accumulation in their bodies is insufficient. Recommended daily energy intake for these infants ranges from 10 to 130 kcal/kg/day to match intrauterine growth rates. Carbohydrates, primarily in the form of glucose, are the main energy source (4 kcal/gram) for newborns. Premature infants exhibit much higher rates of glucose synthesis (6–8 mg/kg/min) compared to full-term neonates (3–5 mg/kg/min) [1,2,11].

Studies have indicated that higher energy and protein intake during the first month after birth in preterm infants are associated with head circumference growth and improved cognitive outcomes in adolescence. Furthermore, enteral calorie supply appears to be more effective than parenteral delivery. For instance, a recent randomized clinical trial comparing early Parenteral Nutrition (PN) with enhanced early PN did not show differences in neurodevelopment but revealed that the enteral intake of calories and proteins in the first week was associated with improved processing speed in evoked potential tests [2,18,25].

The enhanced nutritional protocol entails initiating total parenteral nutrition with a fluid volume of 80 mL/kg/day to provide 4 g/kg/day of amino acids and a minimum glucose infusion rate of 5.5 mg/kg/min. Glucose infusion is increased by 1.5 mg/kg/min daily during the first week, up to a maximum of 12–14 mg/kg/min. Lipids are administered at 2.5–3.5 g/kg/day, and amino acids are maintained at 4 g/kg/day until enteral feeding reaches an average volume of 100 mL/kg/day [2].

Hyperglycemia for more than 5 days in premature infants under 32 weeks of gestational age has been associated with a lower lean mass at 4 months and worse neurodevelopment at 12 months of corrected age. This relationship may be linked to reduced glucose infusion rates in the first week to manage hyperglycemia [26].

5. Lipids

Premature birth is associated with a deficiency of long-chain polyunsaturated fatty acids (LCPUFAs), including docosahexaenoic acid (DHA) and arachidonic acid (ARA). This deficiency persists due to the ineffective conversion of precursor fatty acids, lower fat stores, and limited nutritional supply of DHA and ARA. LCPUFAs are essential for neurodevelopment, normal vision, and protection against complications like Bronchopulmonary Dysplasia, Retinopathy of Prematurity, and Necrotizing Enterocolitis [18,24].

LCPUFAs also play a crucial role in moderating the effects of hypoxia, inflammation, infection, thrombosis, and oxidative damage in key organs like the lungs, brain, and retina. DHA influences the structure of neuronal membranes, synaptogenesis, and myelination. Supplementation with DHA in formula for premature babies has been associated with improved electroretinogram activity, resulting in better visual acuity and overall development in the short term [24,27].

Higher energy and lipid supply in the first two weeks of life have been linked to a reduced risk of severely abnormal MRI findings at full-term age, especially in gray matter, cortex, and cerebellum. Enteral lipids appear to be more effective than intravenous lipid emulsions in this regard. The variable LCPUFA content in different lipid emulsions may account for the differences in outcomes. Additionally, a greater growth of subcortical structures, the cerebellum, and the entire brain, as well as accelerated microstructural maturation of white matter, have been observed in preterm infants under 30 weeks who received more extensive lipid supplementation in enteral nutrition. Enhanced energy and lipid intake may mitigate the adverse impact of respiratory morbidity on brain development, resulting in improved neurodevelopment at 18 months corrected age [5,16,28].

Enteral lipid supplementation has generated conflicting results. While early LCPUFA supplementation in preterm infants has shown positive impacts on early childhood psychomotor neurodevelopment and visual acuity, its effects on long-term global intelligence quotient (IQ) have been less significant [24,29]. Meta-analyses have indicated improved neurodevelopment in premature infants receiving LCPUFA supplementation, as assessed by the Bayley scales between one and three years of age. However, supplementation during lactation accelerates neurodevelopment, with no subsequent changes in developmental outcomes, leading to some frustration among researchers [30–32].

Current recommendations for follow-on formulas suggest that they should contain LCPUFAs at levels similar to those of human milk [33]. It is essential to recognize that nutritional management often fails to provide sufficient preformed DHA during parenteral and enteral nutrition in extremely premature or very low-birth-weight infants due to the need for larger quantities to compensate for intestinal malabsorption, DHA oxidation, and early deficits [34].

The early provision of breast milk, rich in DHA, promotes enhanced brain development in premature infants, supported by substantial evidence. High DHA levels in breast milk have been associated with lower incidence and severity of intraventricular hemorrhage, reduced internal capsule damage, improved white matter development, and better language and motor outcomes at 30–36 months of age. The gender-specific differentiation reveals more significant benefits for boys in terms of brain structure growth and protection against white matter damage [35,36].

6. Human Milk Oligosaccharides (HMOs)

Human milk contains various bioactive components with immunological functions that protect against infections, promote microbial community organization to support organ maturation, and facilitate lactocrine programming, which contributes to favorable neurodevelopmental outcomes [37]. However, a recent narrative review examining associations between exposure to Human Milk Oligosaccharides (HMO) during childhood and neurological development up to 24 months of age found limited evidence to support better neurodevelopment outcomes, particularly among premature infants [37].

Human milk oligosaccharides (HMOs) are the third most abundant solid component of human milk, following lactose and lipids. They are a group of structurally diverse complex carbohydrates, with over 150 distinct structural permutations resulting from HMO biosynthesis. The physiological functions of HMOs that influence brain maturation can vary depending on slight structural differences among them [38].

HMOs can serve as both direct and indirect sources of sialic acid, which is an essential nutrient for the organization of brain tissues. However, it is important to note that the concentration of HMOs in human milk can vary significantly based on the phenotype of the mother and the stage of lactation. Typically, HMO concentrations, relative to their abundance, are higher in colostrum compared to transitional milk and higher in transitional milk compared to mature milk [39].

Some animal studies have identified specific HMOs that may influence early brain maturation. Sialic acids (Sia), both in their free form (N-acetylneuraminic acid) and conjugated forms like 6′-sialyllactose (6′-SL), have been shown to improve cognitive abilities and memory when supplemented in rats. Recent reviews have highlighted the consistent benefits of certain HMOs, such as 2′-FL, 3-FL, 3′-SL, and 6′-SL, for improving motor skills, language development, working memory, and reference memory. However, more research through randomized and controlled clinical trials is needed to fully understand the specific mechanisms involved [40,41].

The volume of formula or human milk consumed by the preterm infant may impact the dose of HMOs, thus affecting neurological development outcomes, that is, greater the accepted volume of enteral feeding with breast milk and its HMO, and better short- and long-term outcomes.

7. Micronutrients

The process of myelination in the developing brain is dependent on various micronutrients, including iron, copper, iodine, vitamin B12, and choline. It is crucial to consider the timing of nutrient supplementation, especially during the critical window of oligodendroglia differentiation, which occurs between 23 and 32 weeks of gestational age when the formation of cerebral white matter is at its peak. Nutritional interventions need to coincide with the development of brain structures or circuits that rely on specific nutrients. Initiating supplementation too early or too late may not produce the desired effects [42].

Iron is essential for the function of enzymes involved in oligodendroglia differentiation and myelination. Iron deficiency is common in premature infants, and it may be associated with impaired maturation and myelination of oligodendrocytes. Studies have shown that iron-deficient neonates often exhibit motor, cognitive, and behavioral delays. Iron deficiency in neonates is often linked to nutritional deficiencies, particularly in premature infants who did not have the opportunity to accumulate sufficient iron stores during the third trimester of pregnancy. Iron's impact on myelin formation has been confirmed through studies of auditory and visual evoked potentials. Delayed umbilical cord clamping in full-term infants has been associated with increased serum ferritin levels and myelin content at four months of age, suggesting potential benefits for myelination [43–46].

Zinc is another crucial mineral for fetal and postnatal development. It plays various roles in gene expression, cell development, replication, and the synthesis of RNA and DNA, which are essential for growth, differentiation, and cellular metabolism. Cerebellar development and NMDA receptor expression also depend on zinc. Extremely preterm infants are particularly vulnerable to zinc deficiency due to diminished stores, increased requirements, and suboptimal absorption. Zinc concentrations in human milk can vary widely. More research is needed to understand the relationship between zinc status, oligodendroglia maturation, myelination, and neurodevelopment in preterm infants [47–49].

8. Vitamins

Vitamin D is known to influence nerve growth factor, promote neurite growth, and inhibit neuronal apoptosis in the hippocampus. A deficiency during critical phases of neurodevelopment can lead to behavioral, memory, and learning disorders later in life. However, a recent systematic review with meta-analysis did not find significant evidence of improved neurodevelopment with enteral vitamin D supplementation for premature and low-birth-weight infants compared to no supplementation or a placebo [50,51].

Data are limited regarding the relationship between selected micronutrients from breast milk, such as vitamin B6, carotenoids, and selenium, and better neurodevelopmental outcomes. It is increasingly believed that a combination of nutritional factors, rather than a single nutrient, has a more significant impact on the neurodevelopment of preterm infants. A study involving British premature boys who received a high-nutrient premature formula in the first 4 weeks of life showed significantly larger volumes of the caudate nucleus at 16 years of age compared to those who received a standard full-term infant formula during the same period, highlighting the potential benefits of a comprehensive nutrient approach [52,53].

9. Intestinal Microbiota and the Gut-Brain Axis

The period between 24 weeks and 52 weeks of post-conceptional age is critical for neural development, characterized by significant neuronal and glial growth in the brain. During this time, the gut-brain axis plays a vital role, with the gut microbiome influencing brain functions and development. Premature infants often have immature gut microbiomes, impacting neurodevelopment. Challenges unique to premature infants include an underdeveloped intestinal barrier that fails to defend against pathogenic bacteria, leading to dysregulated responses that further compromise the immature immune system. Additionally, the underdeveloped blood–brain barrier allows for components associated with the microbiome to cross into the brain more easily, affecting brain functions [1,5,7,54].

HMOs are known to play a crucial role in establishing a healthy microbiota in early life, promoting brain and cognitive development through the gut-brain axis. Potential connections between the intestinal microbiota and the brain include modulation of the immune system, production of neurotransmitters or neuromodulators, regulation of systemic inflammation, and interaction with the Vagus nerve and blood–brain barrier [38,41,54].

Efforts to modulate the microbiome aim to prevent intestinal dysbiosis, which is characterized by an imbalance in microbial colonization. Dysbiosis can result from various exogenous factors, such as mode of delivery, formula feeding, and exposure to antibiotics. Premature infants are particularly susceptible to dysbiosis, which can lead to sepsis and necrotized enterocolitis, both of which are associated with delayed neurodevelopment. Supplementation with probiotics has shown promise in reducing the risk of these conditions and may indirectly contribute to neuroprotection through its trophic effects on the intestine [7,55].

Probiotics have the potential to be neuroprotective due to their direct effects on gene expression, neurotransmitter synthesis, expression of neurotrophic growth factors, and reduction of neuroinflammation. However, the data on the effectiveness of probiotics in premature infants are mixed, with some studies showing benefits in reducing certain risks but not consistently impacting neurocognition [54,56].

10. Human Milk

The advantages of breast milk for neurodevelopment are substantial and indisputable. These benefits are even influenced by the quantity of breast milk consumed. Breast milk contains docosahexaenoic acid (DHA) omega-3 fats, which, when combined with eicosapentaenoic acid (EPA) fats, can potentially reduce the risk of affective disorders, such as major depression and bipolar disorders, ultimately having a positive impact on individuals in society as they mature. While the use of human milk for feeding preterm infants has been shown to offer various neurodevelopmental advantages, other nutritional aspects that affect the growth of premature babies, including macronutrients and micronutrients, may necessitate supplementation to meet their higher nutritional requirements [1,20].

When considering consensus recommendations for feeding preterm infants, it is important to note that the first choice for feeding preterm infants is human milk expressed by their own mothers, with the second option being donated pasteurized human milk. Unpasteurized milk should not be used in the case of human donor milk [3]. The bond between mother and child has a positive association with neurodevelopmental outcomes, and this bond tends to be stronger among breastfeeding mothers, underscoring the importance of policies in neonatal intensive care units (NICUs) that promote the presence of the mother, family-centered care, and breastfeeding.

Belfort et al. observed that preterm infants born before 30 weeks' gestation or with a birth weight less than 1250 g who predominantly received breast milk in the first 28 days of life (more than 50% of their diet) had a larger deep-nuclear-gray matter volume at term-equivalent age. Additionally, they exhibited better IQ, academic achievement, working memory, and motor function at the age of 7. The study focused on regional volumetric measurements obtained through structural magnetic resonance imaging (MRI), a non-invasive technique for studying the brain's anatomy and pathology, distinct from functional magnetic resonance imaging (fMRI), which assesses brain activity [57].

Deoni et al. found that the composition of infant nutrition has an impact on myelin development. Their study revealed that breastfed children showed improved global cognitive ability and rates of cognitive development, including verbal and nonverbal functioning when compared to those receiving infant formula alone. Moreover, exclusive breastfeeding for at least 3 months was associated with enhanced diffuse myelination throughout the brain by the age of 2 [58].

A prior study on activation and connectivity in breastfed infants observed significantly higher gray matter volume in the left and right parietal lobes, as well as the left temporal lobe in breastfed children. Furthermore, breastfed children exhibited greater activation in the right frontal and temporal lobes during perception tasks, while for language tasks, activation was greater in the left temporal lobe [59]. It is well-established that brain activation is positively correlated with performance on cognitive tasks, further reinforcing the evidence that breastfeeding is associated with improved performance on intelligence tests. Table 2 summarizes the advantages of human milk on neurodevelopmental outcomes.

Table 2. Advantages of human milk on neurodevelopment outcome.

Immediate Effects and Short Term	Long Term
Significant improvement in white matter microstructure	Better cognitive, behavioral, and academic performance
Larger deep nuclear gray matter and hippocampus volume at term-equivalent age	Improved working memory
Significantly greater brain volume and white matter volume	Significantly improved verbal IQ, especially in boys (25% increase in IQ)
Better receptive language at age 3	Better verbal and nonverbal IQ at age 7 *
Improved mental and psychomotor development scores at Bayley Scales	Significantly higher IQ in later years, even after adjustment for maternal IQ

* This result was observed in a longer duration of breast milk feeding; each month of breastfeeding can increase verbal IQ by 0.35 points and nonverbal IQ by 0.29 points. Adapted from Kumar et al. 2017 [3] and De Nardo et al. 2022 [5].

Despite numerous studies, the precise mechanisms through which the components of human milk shape the structural and functional characteristics of the infant brain remain incompletely understood. This area of research holds great promise, as children born prematurely who receive a greater supply of their own mother's breast milk appear to exhibit higher general intelligence, improved academic performance, enhanced memory, and better motor function as they grow [16,38,41,60]. Thus, breast milk plays a pivotal role in promoting overall brain development, emerging as a key factor contributing to the positive effects of breastfeeding on intelligence.

11. Conclusions

Nutrition is a critical factor in achieving adequate growth and brain development in premature infants, especially those with lower gestational ages. While nutrition alone may not fully counteract the challenges of extreme prematurity, nutritional therapies hold promise for promoting brain development. These interventions, including amino acids, glucose, and LCPUFAs, are considered safe, cost-effective, and minimally invasive.

They can be combined with breastfeeding, which is the optimal source of nutrition for premature infants.

In addition to the emphasis on macronutrient intake, it is crucial to consider micronutrients, neuropeptides, neurohormones, and their roles in modulating the gut-immune-brain axis. Further research is needed to explore the potential impact of donated human milk and milk pasteurization on the premature infant's brain. Individualized nutritional approaches that address the unique trajectories of premature infants and the evolving nutritional needs are essential.

Breastfeeding may improve maternal–infant bonding, which is another possible mechanism for the positive breastfeeding association with development.

Overall, the growing recognition of human milk as the gold standard of nutrition, along with the integration of comprehensive nutrient strategies, provides hope for better outcomes in the neurodevelopment of premature infants.

Funding: This research received no external funding.

Conflicts of Interest: The authors declare no conflict of interest.

References

1. Ramel, S.E.; Georfieff, M.K. Preterm nutrition and the brain. In *Nutritional Care of Preterm Infants*; Koletzko, B., Poindexter, B., Uauy, R., Eds.; Karger: Basel, Switzerland, 2014; pp. 190–200.
2. Morris, E.E.; Miller, N.C.; Marka, N.A.; Super, J.L.; Nagel, E.M.; Gonzalez, J.D.; Demerath, E.W.; Ramel, S.E. Randomized Trial of Early Enhanced Parenteral Nutrition and Later Neurodevelopment in Preterm Infants. *Nutrients* **2022**, *14*, 3890. [CrossRef] [PubMed]
3. Kumar, R.K.; Singhal, A.; Vaidya, U.; Banerjee, S.; Anwar, F.; Rao, S. Optimizing Nutrition in Preterm Low Birth Weight Infants—Consensus Summary. *Front. Nutr.* **2017**, *4*, 20. [CrossRef] [PubMed]
4. Johnson, M.J.; Wootton, S.A.; Leaf, A.A.; Jackson, A.A. Preterm birth and body composition at term equivalent age: A systematic review and meta-analysis. *Pediatrics* **2012**, *130*, e640–e649. [CrossRef] [PubMed]
5. De Nardo, M.C.; Di Mario, C.; Laccetta, G.; Boscarino, G.; Terrin, G. Enteral and parenteral energy intake and neurodevelopment in preterm infants: A systematic review. *Nutrition* **2022**, *97*, 111572. [CrossRef] [PubMed]
6. Samara, J.; Moossavi, S.; Alshaikh, B.; Ortega, V.A.; Pettersen, V.K.; Ferdous, T.; Hoops, S.L.; Soraisham, A.; Vayalumkal, J.; Dersch-Mills, D.; et al. Supplementation with a probiotic mixture accelerates gut microbiome maturation and reduces intestinal inflammation in extremely preterm infants. *Cell Host Microbe* **2022**, *30*, 696–711.e5. [CrossRef]
7. Buffet-Bataillon, S.; Bellanger, A.; Boudry, G.; Gangneux, J.-P.; Yverneau, M.; Beuchée, A.; Blat, S.; Le Huërou-Luron, I. New Insights Into Microbiota Modulation-Based Nutritional Interventions for Neurodevelopmental Outcomes in Preterm Infants. *Front. Microbiol.* **2021**, *12*, 676622. [CrossRef]
8. Beghetti, I.; Barone, M.; Turroni, S.; Biagi, E.; Sansavini, A.; Brigidi, P.; Corvaglia, L.; Aceti, A. Early-life gut microbiota and neurodevelopment in preterm infants: Any role for Bifidobacterium? *Eur. J. Pediatr.* **2022**, *181*, 1773–1777. [CrossRef] [PubMed]
9. Cooke, R.W. Preterm mortality and morbidity over 25 years. *Arch. Dis. Child Fetal Neonatal Ed.* **2006**, *91*, F293–F294. [CrossRef]
10. Sandström, O.; Lönnerdal, B.; Graverholt, G.; Hernell, O. Effects of alpha-lactalbumin-enriched formula containing different concentrations of glycomacropeptide on infant nutrition. *Am. J. Clin. Nutr.* **2008**, *87*, 921–928. [CrossRef]
11. Ehrenkranz, R.A. Early, Aggressive nutritional management for very low birth weight infants: What is the evidence? *Semin. Perinatol.* **2007**, *31*, 48–55. [CrossRef]
12. Morgan, C.; McGowan, P.; Herwitker, S.; Hart, A.E.; Turner, M.A. Postnatal head growth in preterm infants: A randomized controlled parenteral nutrition study. *Pediatrics* **2014**, *133*, e120–e128. [CrossRef]
13. Leppänen, M.; Lapinleimu, H.; Lind, A.; Matomäki, J.; Lehtonen, L.; Haataja, L.; Rautava, P.; on Behalf of the PIPARI Study Group. Antenatal and postnatal growth and 5-year cognitive outcome in very preterm infants. *Pediatrics* **2014**, *133*, 63–70. [CrossRef]
14. Cormack, B.E.; Bloomfield, F.H.; Dezoete, A.; Kuschel, C.A. Does more protein in the first week of life change outcomes for very low birthweight babies? *J. Paediatr. Child Health* **2011**, *47*, 898–903. [CrossRef] [PubMed]
15. Keunen, K.; van Elburg, R.M.; van Bel, F.; Benders, M.J.N.L. Impact of nutrition on brain development and its neuroprotective implications following preterm birth. *Pediatr. Res.* **2015**, *77*, 148–155. [CrossRef] [PubMed]
16. Schneider, N.; Garcia-Rodenas, C.L. Early Nutritional Interventions for Brain and Cognitive Development in Preterm Infants: A Review of the Literature. *Nutrients* **2017**, *9*, 187. [CrossRef]
17. Roberts, D.; Brown, J.; Medley, N.; Dalziel, S.R. Antenatal corticosteroids for accelerating fetal lung maturation for women at risk of preterm birth. *Cochrane Database Syst. Rev.* **2017**, *3*, CD004454. [CrossRef] [PubMed]
18. Robinson, D.T.; Calkins, K.L.; Chen, Y.; Cober, M.P.; Falciglia, G.H.; Church, D.D.; Mey, J.; McKeever, L.; Sentongo, T. Guidelines for parenteral nutrition in preterm infants: The American Society for Parenteral and Enteral Nutrition. *J. Parenter. Enter. Nutr.* **2023**, *47*, 830–858. [CrossRef]

19. Fenton, T.R.; Al-Wassia, H.; Premji, S.S.; Sauve, R.S. Higher versus lower protein intake in formula-fed low birth weight infants. *Cochrane Database Syst. Rev.* **2020**, *6*, CD003959. [CrossRef]
20. Gao, C.; Miller, J.; Collins, C.T.; Rumbold, A.R. Comparison of different protein concentrations of human milk fortifier for promoting growth and neurological development in preterm infants. *Cochrane Database Syst. Rev.* **2020**, *11*, CD007090. [CrossRef]
21. Embleton, N.D.; Akker, C.H.v.D. Protein intakes to optimize outcomes for preterm infants. *Semin. Perinatol.* **2019**, *43*, 151154. [CrossRef]
22. Wharton, B.A.; Morley, R.; Isaacs, E.B.; Cole, T.J.; Lucas, A. Low plasma taurine and later neurodevelopment. *Arch. Dis. Child.-Fetal Neonatal Ed.* **2004**, *89*, F497–F498. [CrossRef] [PubMed]
23. Volpe, J.J. *Volpe's Neurology of the Newborn*; Elsevier: Amsterdam, The Netherlands, 2017; pp. 100–177.
24. Cormack, B.E.; Harding, J.E.; Miller, S.P.; Bloomfield, F.H. The Influence of Early Nutrition on Brain Growth and Neurodevelopment in Extremely Preterm Babies: A Narrative Review. *Nutrients* **2019**, *11*, 2029. [CrossRef] [PubMed]
25. Isaacs, E.B.; Morley, R.; Lucas, A. Early diet and general cognitive outcome at adolescence in children born at or below 30 weeks gestation. *J. Pediatr.* **2009**, *155*, 229–234. [CrossRef] [PubMed]
26. Villamizar, J.D.G.; Haapala, J.L.; Scheurer, J.M.; Rao, R.; Ramel, S.E. Relationships between Early Nutrition, Illness, and Later Outcomes among Infants Born Preterm with Hyperglycemia. *J. Pediatr.* **2020**, *223*, 29–33.e2. [CrossRef] [PubMed]
27. Georgieff, M.K. Nutrition and the developing brain: Nutrient priorities and measurement. *Am. J. Clin. Nutr.* **2007**, *85*, 614S–620S.
28. Skinner, A.M.; Narchi, H. Preterm nutrition and neurodevelopmental outcomes. *World J. Methodol.* **2021**, *11*, 278–293. [CrossRef]
29. Shulkin, M.; Pimpin, L.; Bellinger, D.; Kranz, S.; Fawzi, W.; Duggan, C.; Mozaffarian, D. n–3 Fatty Acid Supplementation in Mothers, Preterm Infants, and Term Infants and Childhood Psychomotor and Visual Development: A Systematic Review and Meta-Analysis. *J. Nutr.* **2018**, *148*, 409–418. [CrossRef]
30. Wang, Q.; Cui, Q.; Yan, C. The Effect of Supplementation of Long-Chain Polyunsaturated Fatty Acids During Lactation on Neurodevelopmental Outcomes of Preterm Infant From Infancy to School Age: A Systematic Review and Meta-analysis. *Pediatr. Neurol.* **2016**, *59*, 54–61.e1. [CrossRef]
31. Alshweki, A.; Muñuzuri, A.P.; Baña, A.M.; de Castro, M.J.; Andrade, F.; Aldamiz-Echevarría, L.; de Pipaón, M.S.; Fraga, J.M.; Couce, M.L. Effects of different arachidonic acid supplementation on psychomotor development in very preterm infants; a randomized controlled trial. *Nutr. J.* **2015**, *14*, 101. [CrossRef]
32. Moon, K.; Rao, S.C.; Schulzke, S.M.; Patole, S.K.; Simmer, K. Long chain polyunsaturated fatty acid supplementation in preterm infants. *Cochrane Database Syst. Rev.* **2016**, *12*, CD000375.
33. Koletzko, B.; Bergmann, K.; Brenna, J.T.; Calder, P.C.; Campoy, C.; Clandinin, M.T.; Colombo, J.; Daly, M.; Decsi, T.; Demmelmair, H.; et al. Should formula for infants provide arachidonic acid along with DHA? *Am. J. Clin. Nutr.* **2020**, *111*, 10–16. [CrossRef] [PubMed]
34. Lapillonne, A.; Groh-Wargo, S.; Gonzalez, C.H.L.; Uauy, R. Lipid Needs of Preterm Infants: Updated Recommendations. *J. Pediatr.* **2013**, *162* (Suppl. 3), S37–S47. [CrossRef]
35. Tam, E.W.Y.; Chau, V.; Barkovich, A.J.; Ferriero, D.M.; Miller, S.P.; Rogers, E.E.; Grunau, R.E.; Synnes, A.R.; Xu, D.; Foong, J.; et al. Early postnatal docosahexaenoic acid levels and improved preterm brain development. *Pediatr. Res.* **2016**, *79*, 723–730. [CrossRef] [PubMed]
36. Ottolini, K.M.; Andescavage, N.; Keller, S.; Limperopoulos, C. Nutrition and the developing brain: The road to optimizing early neurodevelopment: A systematic review. *Pediatr. Res.* **2020**, *87*, 194–201. [CrossRef] [PubMed]
37. Berger, P.K.; Ong, M.L.; Bode, L.; Belfort, M.B. Human Milk Oligosaccharides and Infant Neurodevelopment: A Narrative Review. *Nutrients* **2023**, *15*, 719. [CrossRef]
38. Bode, L.; Jantscher-Krenn, E. Structure-function relationships of human milk oligosaccharides. *Adv. Nutr. Int. Rev. J.* **2012**, *3*, 383S–391S. [CrossRef]
39. Azad, M.B.; Robertson, B.; Atakora, F.; Becker, A.B.; Subbarao, P.; Moraes, T.J.; Mandhane, P.J.; Turvey, S.E.; Lefebvre, D.L.; Sears, M.R.; et al. Human Milk Oligosaccharide Concentrations Are Associated with Multiple Fixed and Modifiable Maternal Characteristics, Environmental Factors, and Feeding Practices. *J. Nutr.* **2018**, *148*, 1733–1742. [CrossRef]
40. Oliveros, E.; Vázquez, E.; Barranco, A.; Ramírez, M.; Gruart, A.; Delgado-García, J.M.; Buck, R.; Rueda, R.; Martín, M.J. Sialic Acid and Sialylated Oligosaccharide Supplementation during Lactation Improves Learning and Memory in Rats. *Nutrients* **2018**, *10*, 1519. [CrossRef] [PubMed]
41. Fan, Y.; McMath, A.L.; Donovan, S.M. Review on the Impact of Milk Oligosaccharides on the Brain and Neurocognitive Development in Early Life. *Nutrients* **2023**, *15*, 3743. [CrossRef]
42. Georgieff, M.K.; Ramel, S.E.; Cusick, S.E. Nutritional influences on brain development. *Acta Paediatr.* **2018**, *107*, 1310–1321. [CrossRef]
43. Berglund, S.K.; Westrup, B.; Hägglöf, B.; Hernell, O.; Domellöf, M. Effects of iron supplementation of LBW infants on cognition and behavior at 3 years. *Pediatrics* **2013**, *131*, 47–55. [CrossRef] [PubMed]
44. Berglund, S.K.; Chmielewska, A.; Starnberg, J.; Westrup, B.; Hägglöf, B.; Norman, M.; Domellöf, M. Effects of iron supplementation of low-birth-weight infants on cognition and behavior at 7 years: A randomized controlled trial. *Pediatr. Res.* **2017**, *83*, 111–118. [CrossRef]
45. Volpe, J.J. Iron and zinc: Nutrients with potential for neurorestoration in premature infants with cerebral white matter injury. *J. Neonatal-Perinatal Med.* **2020**, *12*, 365–368. [CrossRef]

46. Rao, R.; Bora, R. Timing of umbilical cord clamping and infant brain development. *J. Pediatr.* **2018**, *203*, 8–10. [CrossRef] [PubMed]
47. Brion, L.P.; Heyne, R.; Lair, C.S. Role of zinc in neonatal growth and brain growth: Review and scoping review. *Pediatr. Res.* **2021**, *89*, 1627–1640. [CrossRef] [PubMed]
48. Castillo-Durán, C.; Weisstaub, G. Zinc Supplementation and growth of the fetus and low birth weight infant. *J. Nutr.* **2003**, *133*, 1494S–1497S. [CrossRef]
49. Griffin, I.J.; Domellöf, M.; Bhatia, J.; Anderson, D.M.; Kler, N. Zinc and copper requirements in preterm infants: An examination of the current literature. *Early Hum. Dev.* **2013**, *89*, S29–S34. [CrossRef] [PubMed]
50. Koletzko, B.; Cheah, F.C.; Domellöf, M.; Poindexter, B.B.; Vain, N.; van Goudoever, J.B. Nutritional Care of Preterm Infants. Scientific Basis and Practical Guidelines. *World Rev. Nutr. Diet.* **2021**, *122*, 46–59. [CrossRef]
51. Kumar, M.; Shaikh, S.; Sinha, B.; Upadhyay, R.P.; Choudhary, T.S.; Chandola, T.R.; Mazumder, S.; Taneja, S.; Bhandari, N.; Chowdhury, R. Enteral Vitamin D Supplementation in Preterm or Low Birth Weight Infants: A Systematic Review and Meta-analysis. *Pediatrics* **2022**, *150*, 150–160. [CrossRef] [PubMed]
52. Lockyer, F.; McCann, S.; Moore, S.E. Breast Milk Micronutrients and Infant Neurodevelopmental Outcomes: A Systematic Review. *Nutrients* **2021**, *13*, 3848. [CrossRef] [PubMed]
53. Isaacs, E.B.; Gadian, D.G.; Sabatini, S.; Chong, W.K.; Quinn, B.T.; Fischl, B.R.; Lucas, A. The effect of early human diet on caudate volumes and IQ. *Pediatr. Res.* **2008**, *63*, 308–314. [CrossRef] [PubMed]
54. Lu, J.; Claud, E.C. Connection between gut microbiome and brain development in preterm infants. *Dev. Psychobiol.* **2019**, *61*, 739–751. [CrossRef]
55. Arboleya, S.; Sánchez, B.; Milani, C.; Duranti, S.; Solís, G.; Fernández, N.; de Los Reyes-Gavilan, C.G.; Ventura, M.; Margolles, A.; Gueimonde, M. Intestinal microbiota development in preterm neonates and effect of perinatal antibiotics. *J. Pediatr.* **2015**, *166*, 538–544. [CrossRef] [PubMed]
56. Panchal, H.; Athalye-Jape, G.; Rao, S.; Patole, S. Growth and neuro-developmental outcomes of probiotic supplemented preterm infants—A systematic review and meta-analysis. *Eur. J. Clin. Nutr.* **2023**, *77*, 855–871. [CrossRef] [PubMed]
57. Belfort, M.B.; Anderson, P.J.; Nowak, V.A.; Lee, K.J.; Molesworth, C.; Thompson, D.K.; Doyle, L.W.; Inder, T.E. Breast Milk Feeding, Brain Development, and Neurocognitive Outcomes: A 7-Year Longitudinal Study in Infants Born at Less Than 30 Weeks' Gestation. *J. Pediatr.* **2016**, *177*, 133–139.e1. [CrossRef] [PubMed]
58. Deoni, S.; Dean, D.; Joelson, S.; O'Regan, J.; Schneider, N. Early nutrition influences developmental myelination and cognition in infants and young children. *NeuroImage* **2017**, *178*, 649–659. [CrossRef] [PubMed]
59. Ou, X.; Andres, A.; Pivik, R.; Cleves, M.; Snow, J.; Ding, Z.; Badger, T. Voxel-Based Morphometry and fMRI Revealed Differences in Brain Gray Matter in Breastfed and Milk Formula–Fed Children. *Am. J. Neuroradiol.* **2016**, *37*, 713–719. [CrossRef]
60. Boscarino, G.; Di Chiara, M.; Cellitti, R.; De Nardo, M.C.; Conti, M.G.; Parisi, P.; Spalice, A.; Di Mario, C.; Ronchi, B.; Russo, A.; et al. Effects of early energy intake on neonatal cerebral growth of preterm newborn: An observational study. *Sci. Rep.* **2021**, *11*, 18457. [CrossRef]

MDPI AG
Grosspeteranlage 5
4052 Basel
Switzerland
Tel.: +41 61 683 77 34

Nutrients Editorial Office
E-mail: nutrients@mdpi.com
www.mdpi.com/journal/nutrients

www.ingramcontent.com/pod-product-compliance
Lightning Source LLC
Chambersburg PA
CBHW042009260926
51850CB00020B/2942
* 9 7 8 3 7 2 5 8 4 4 4 7 0 *